RECOVERING THE SOUL

BANTAM NEW AGE BOOKS

RECOVERING THE SOUL

A Scientific and Spiritual Search

Larry Dossey, M.D.

BANTAM BOOKS

NEW YORK • TORONTO • LONDON • SYDNEY • AUCKLAND

Many of the clinical cases in this book are based on actual events occurring in the author's practice of internal medicine. All names have been changed to preserve confidentiality.

RECOVERING THE SOUL
A Bantam Book / December 1989

Library of Congress Cataloging-in-Publication Data

Dossey, Larry, 1940–
Recovering the soul : a scientific and spiritual search / Larry Dossey.
p. cm.
Bibliography: p.
Includes index.
ISBN 0-553-34790-X
1. Mind and body. 2. Consciousness. 3. Soul. 4. Transpersonal psychology. I. Title.
BF161.D68 1989
153—dc20 89-6721
CIP

Published simultaneously in the United States and Canada

Bantam Books are published by Bantam Books, a division of Bantam Doubleday Dell Publishing Group, Inc. Its trademark, consisting of the words "Bantam Books" and the portrayal of a rooster, is Registered in U.S. Patent and Trademark Office and in other countries. Marca Registrada. Bantam Books, 666 Fifth Avenue, New York, New York 10103.

PRINTED IN THE UNITED STATES OF AMERICA

O 0 9 8 7 6 5 4 3 2 1

For my parents

Acknowledgments

I am especially grateful to The Isthmus Institute of Dallas, Texas, where an ongoing dialogue about the possible convergences of science and religious thought has provided inspiration for a decade.

I would like to thank my patients, who are the source of most of the clinical material that follows, as well as my colleagues at Dallas Diagnostic Association for their kind support.

The persons who have contributed to the evolution of this book are too numerous to mention. If there is truth to the ideas that follow, perhaps a collective "thank you" will be received by each of them, via the One Mind that connects us all.

Thanks are due literary agents John Brockman and Katinka Matson for finding a home for the publication of this book, and especially to Leslie Meredith of Bantam, my remarkable editor, whose editorial creativity is exceeded only by her kindness and understanding of the material.

But most of all, thanks go to Barbara for enabling this work to come to be in endless ways—fellow writer, nurse, colleague, and wife.

LARRY DOSSEY, M.D.

Contents

Part II—SCIENCE: The Proofs

Part III—GOD: The Synthesis

The old gods are dead or dying
and people everywhere are searching,
asking: What is the new mythology to be,
the mythology of this unified earth
as one harmonious being?

—Joseph Campbell

RECOVERING THE SOUL

Preface:

"Eternal, Infinite, and One"

Why speak of the soul in an age of science? Why suggest that there is some aspect of the psyche that is not subject to the limitations of space and time, and which might precede the birth of the body and survive its death? The main reason is that something vital has been left out of almost all the modern efforts to understand our mental life—something that counts as a first principle, without which everything is bound to be incomplete and off base.

This missing element is the mind's *nonlocal* nature. Virtually every modern explanation of the mind assumes that it can be found in a certain place in space or time. The spatial place to which the mind has been traditionally fixed is the brain of an individual person; the temporal place to which it is assigned is the present moment and a single lifetime. Thus, a mind is consigned and isolated to an individual person. And, what is worse, it is thus doomed to oblivion, for with the death of the brain it must also die. These local assumptions underlie almost all the scholarly Western attempts to describe the human mind. Because they try to *locate* the mind, to fix it in time and space, they are, I believe, flawed.

As we shall see, there is good evidence that the mind cannot be localized. It displays its nonlocal character in a million ways, showing us that it is free in space and time, that it bridges consciousness between persons, and that it does not die with the body. This book explores some alternative ways of viewing the mind, and in particular gives scientific evidence for a broader vision of it.

If there is an aspect of the mind that is indeed nonlocal, then this entity comes to resemble the soul—something that is timeless, spaceless, and immortal. *Recovering the nonlocal nature of the mind, then, is essentially a recovery of the soul.* That is why there is such great spiritual significance to understanding the nonlocal nature of our mental life. Unraveling the nonlocal nature of the mind is not just an exercise in psychology or science; it is an exercise in spirituality as well, because of the soul-like nature of a mind that is nonlocal.

"Mind," "soul," and "consciousness" are all dealt with in this inquiry. What do these different words mean? Alas, it is not possible to separate these concepts in a universally acceptable way. Biologists, information theorists, artificial intelligence experts, and other scientists take many different positions, which usually conflict with those of philosophers, psychologists, and theologians. And the differences are not trivial. For some, "mind" simply does not exist. It is a "ghost in the machine," a redundant and unnecessary abstraction of events that can be fully explained in uninflated, physical terms. For others, "mind" is a "category mistake," a "confusion of explanatory levels," or an epiphenomenon or by-product of chemical actions within the brain. Still others regard it as an entity that is quite real.

One soon feels that attempts to define mind, soul, and consciousness, if not futile, are like the fable of the blind men describing an elephant: each description depends on the part of the anatomy the describer happens to be holding at the time.

That does not imply, though, that meanings do not exist or that mind and consciousness are illusions. The elephant is real enough. It may indicate, however, that definitions of the type employed in empirical science cannot be given.

In this book I shall follow physicist and philosopher Henry Margenau and regard mind and consciousness as "primitives"—things indefinable in terms of empirical facts. This choice, Margenau contends, "is justified not only by the rules of logic but is made cogent by the fact that consciousness is at once the most immediate personal experience and the source from which all knowledge springs." [1] For Margenau and this inquiry, mind equals consciousness. Furthermore, for reasons to be given, these entities are considered nonmaterial, infinite in space, eternal, unconfined to brains and bodies, and capable of exerting change in the physical world. Sometimes, when "mind" is used in this larger sense, I will refer to it as Mind.

Although Mind is neither confined to the brain nor a product of it, Mind may nonetheless work through the brain. The result is the appearance of individual minds, derivative of the larger Mind, which we refer to as the individual self, the ego, the person, and the sense of I. The primary characteristics of minds are *content* and some level of conscious awareness: the myriad thoughts, emotions, and sensations that flood us daily. Individual minds are highly susceptible to changes in the physical body: moods, emotions, and even thoughts can be modified by changes in the brain and body.

There are many levels of consciousness—a "spectrum," to use psychologist Ken Wilber's term, that includes those levels commonly acknowledged in the West such as the unconscious, the preconscious, and the conscious. In addition, there are higher levels that have long been recognized in the elegant typologies of the East, but which are rarely spoken of in our culture. Mahayana Buddhism, Taoism, Sufism, Vedanta, Vajrayana, and various schools of yoga are some of the systems that describe a span reaching all the way from the primitive, undifferentiated, preeogic state of the infant to the level of ultimate unity with the Divine, with many stages in between. In the West we have stopped short in describing the span of consciousness at the level of the "healthy, functioning ego," ignoring any higher dimensions that may exist. A remarkable exception is the work of Wilber who, in his landmark book, *Spectrum of Consciousness,* has fashioned a model resembling those of the East. [2] Lying above the level of the "healthy ego" are several other strata—various "subtle" and "causal" realms, and finally the state of "ultimate unity." I will frequently refer to the latter state of ultimate oneness as the highest Self, the Soul, and the One Mind, which contain attributes of the Divine. As this principle is frequently stated in Western religions, the "home" of the soul is God; in the East, Atman (the individual soul) and Brahman (the Ultimate) are one.

A principle embodied in many spiritual traditions that is helpful in keeping these distinctions straight is: "The higher contains the lower, but the lower does not contain the higher." Just so, Mind contains minds, Self contains selves, and Soul contains souls. But the ultimate entities—Mind, Consciousness, and Soul—cannot be explained by, nor are they fully contained in, the lesser entities.

Still, these terms are not always used with precision in this book, nor, perhaps, should they be. Scientist and philosopher Lewis Thomas has suggested that language, if it is to be effective, must be allowed to stray off course. It is wrong, he insists, always to demand that words follow a rigid, locked-on trajectory. Although this view can obviously be used as an excuse for sloppiness and imprecision, communication must paradoxically be allowed a certain ambiguity. Thus the meanings of mind, soul, and self may drift a bit in this book. But the meanings, I trust, can always be gleaned from the context in which they are used. For this I ask the reader's kind indulgence.

The data used in this book to talk about the mind range from poetry to physics, from mysticism to medicine. The reason I have strayed from the hard data of science in the hope of coming up with the most accurate picture of the mind is important. All investigations must deal with "observables" of some sort—observations drawn from experience, out of which some view is fashioned. *There are many possible kinds of observables.* Those of the psychologist are not those of the physicist, and they differ too from those of the poet or mystic. Observables are not equally accessible to all persons. The psychologist does not have the tools or the knowledge with which to observe subatomic behavior; neither does the physicist have the requisite skills to be a preeminent observer of human behavior. Data from the sciences are important, but hardly comprehensive, especially in the domain of "primitives": soul, mind, and consciousness. Thus one may unapologetically choose among many sets of observables.

Today, even the simplest questions about the nature of mind, soul, and consciousness involve difficulties that philosophers of the past were spared. Where exactly, for example, is the mind located? Today physicists think of space and time as a sort of multidimensional sheet called space-time, perhaps connected with other multidimensional sheets. Could the human mind be embedded in, or enfolded in, a higher dimensional space-time, as well as in the one we inhabit? From this higher vantage point, physicist Paul Davies has suggested, the Mind or Soul might "lock on" to the body of an individual in our space-time, without itself being part of our immediate reality. It could be here and there at the same time—perhaps in several different "realities" simultaneously. Moreover, some physicists believe space-time is not fundamental, but derived—built up not of places or moments but of abstract subunits. A few of these could come together to form the space-time we know, leaving an immense ocean of space-time elsewhere in the form of disconnected bits. These bits would be in no particular space, so that assigning a definite location to the mind would be meaningless. [3] And time, Davies reminds us, is not the unchanging entity it is frequently assumed to be; it can be altered physically by human manipulation. If we place the mind in time, does this mean we can affect how long it endures? [4] Even more bewildering, some physicists, as we shall later see, believe actions performed in the present can affect the past, even though the present had not then come into existence! [5]

How can the answers we give to space-and-time questions about God, Mind, Self, and Soul possibly take all these complex ideas into consideration? And even if we wish to be consistent with the way modern physicists view space and time, there is no final verdict within physics about what these terms mean, or what new interpretations may arise. Moreover, as Davies points out, physicists themselves are not consistent in how they employ these concepts. They may use one set of definitions in the laboratory but lapse back into the ideas of common sense when they leave the job. In spite of these uncertainties, I will refer in this book to two types of time: (1) the time of common sense (linear, flowing, external time; the time of progress, development, and history), and (2) the time that is alluded to in modern physics (nonflowing, nonlinear time; the "time of eternity"; the time in which things do not happen, but simply "are").

The following account, then, intentionally paints with a very broad brush. It is not a complete look at the nonlocal nature of the mind. It places the topic on the table, where it awaits further refinement.

But there is a greater reason to explore the nonlocal nature of the mind than simply to "be accurate" in some logical or scientific sense. This reason is conveyed by the Nobel neurophysiologist Sir John Eccles:

> Man has lost his way ideologically in this age. . . . Science has gone too far in breaking down man's belief in his spiritual greatness . . . and has given him the belief that he is merely an insignificant animal that has arisen by chance and necessity in an insignificant planet lost in the great cosmic immensity. . . . We must realize the great unknowns in the material makeup and operation of our brains, and in the relationship of brain to mind and in our creative imagination. [6]

The main reason to establish the nonlocal nature of the mind is, then, spiritual. Local theories of the mind are not only incomplete, they are destructive. They create the illusion of death and aloneness, altogether local concepts. They foster existential oppression and hopelessness by giving us an utterly false idea of our basic nature, advising us that we are contracted, limited, and mortal creatures locked inside our bodies and drifting inexorably toward the end of time. This local scenario is ghastly, and it is regrettable that it continues to dominate the picture put forward by most of our best psychologists and bioscientists.

This book tries to show why this is a false view. If it is false, and we are *non*local instead of local creatures, then the world changes for us in the most glorious ways. For if the mind is nonlocal, it must in some sense be independent of the strictly local brain and body. This opens up the possibility, at least, for some measure of freedom of the will, since the mind could escape the determinative constraints of the physical laws governing the physical body. And if the mind is nonlocal, unconfined to brains and bodies and thus not entirely dependent on the physical organism, the possibility for survival of bodily death is opened. Then there is the nature of our relationship to each other. If the mind is nonlocal in space and time, our interaction with each other seems a foregone conclusion. Nonlocal minds are *merging* minds, since they are not "things" that can be walled off and confined to moments in time or point-positions in space.

If nonlocal mind is a reality, the world becomes a place of interaction and connection, not one of isolation and disjunction. And if humanity really *believed* that nonlocal mind were real, an entirely new foundation for ethical and moral behavior would enter, which would hold at least the possibility of a radical departure from the insane ways human beings and nation-states have chronically behaved toward each other. And, further, the entire existential premise of human life might shift toward the moral and the ethical, toward the spiritual and the holy. Nonlocal mind potentially leads, to borrow historian and sociologist Morris Berman's provocative phrase, to a *reenchantment of the world*.

Suppose for the moment that we *could* show that the human mind is nonlocal; that it is ultimately independent of the physical brain and body and that, as a correlate, it transcends time and space. This, I believe, would rank in importance far beyond anything ever discovered, past or present, about the human organism. This discovery would strike a chord of hope about our inner nature that has been silenced in an age of science; it would stir a new vision of the human as triumphant over flesh and blood: It would anchor the human spirit once again on the side of God instead of randomness, chance, and decay. It would spur the human will to greatness instead of expediency and self-service; it would assuage the bad conscience modern men feel when they dream of innate purposes and goals of life, to say nothing of immortality. With one sweep, this discovery would redirect the imperatives of medicine. No longer would it be the ultimate goal of the modern healer to forestall death and decay, for these would lose their absolute status if the mind were ultimately transcendent over the physical body. The mad, frenzied, life-at-any-cost dictum that prevails today could be modulated in its intensity, along with the despair that dying men feel.

And once again we might recover something that has been notably absent in our experience of late: the human soul.

Inflated stakes? I do not think so. There is reason to set our goals high today, higher than we could have envisioned at any point in the past two hundred years when the physical sciences began to suggest that goals, values, purposes, and the transcendent are not contained in nature.

This book looks to these heights. And it tries to reach them by exploring that single fact which, more than any other I can think of, has the capacity to redirect our vision and restore our ability once again to feel life: the absolute status of human consciousness—consciousness as fundamental and not derivative of the physical; consciousness as infinite in space and time.

In spite of the evidence offered for it, some persons will object in principle to the concept of nonlocal mind. I have discovered that many of these objections seem to be based not on reason but on what could be called "spiritual agoraphobia." Like their counterparts who suffer from the *psychological* affliction of agoraphobia—the fear of open places—*spiritual* agoraphobics have a deep-seated fear of vast expanses: the infiniteness in time and space suggested by nonlocal mind. They feel safer when things are closed in, finite, and "local"—such as a mind that is confined to the individual brain and body, and a mind that stays put in the here-and-now. A mind, in other words, that is soulless.

Some readers may also detect a thread of pantheism in the chapters that follow, especially in the sections suggesting that animals, lower life forms, and humans may all participate in the unbounded One Mind. Since "pantheism" was coined in 1705 by John Toland, the English deist and philosopher, it has frequently been used to denounce attempts that would debase God and eliminate his personality by diffusing divinity, willy nilly, into the world. Although pantheism is still largely a dirty word in the Christian West, perhaps there are reasons why we should not revile it today as much as in former times. We need not recoil from the idea of a God-permeated nature so long as we apply the concept of *hierarchy*—the notion of a multi-tiered world in which there are degrees of Divinity distributed through the world. We are close to such a concept already; today the physical sciences are alive with the idea of "levels of complexity," wherein different properties inhere both in living and nonliving entities, depending on their degree of organizational intricacy. Thus, although one attributes consciousness or mind to the entire material world, one is not required to attribute mind *in the same degree* to, say, both electrons and humans. The differences in the two may remain if the concept of hierarchy is preserved. Just so, we can escape the pitfalls of pantheism and avoid homogenizing God and the world into some unrecognizable blur.

It could be argued that pantheism, far from being a pejorative idea, might be a valuable commodity in today's world. We have been assured by science that the world and all in it are nothing but a dead, unthinking clockwork. Nature, pursuing only the blind laws of physics, contains no meaning, let alone Divinity; any Godlike qualities we see in nature come from us, not from nature itself, and are projections. Freedom of the will and meaningful choice are swamped by the restrictions of physical law. "Mind" is "a ghost in the machine," merely a fanciful term to describe the blind play of atoms in the brain. Moreover, our psychic life is dominated by dark, unconscious forces lying outside our awareness, that push us in directions we seldom comprehend. As if these desultory views from science and psychology were not enough, we are assured from the pulpits that we are sinful, fallen creatures whose only hope for redemption is through the generosity of God. The net result of these preachments from both science and religion has been a near-total disenchantment of the world and all in it, including ourselves—a vulgarization of creation in which any sense of sacredness has been sacrificed.

In contrast to this dismal picture, pantheism seems almost refreshing. It is not necessary to protect the Divine from being contaminated by His world, as those who protest against pantheism seem to think. Rather, the modern challenge is the other way around—to once again find sacredness within the world; to recover our own soul, and the soul of the world.

After all, the question today is not whether humans, animals, plants, and electrons share mind or consciousness equally, or whether or not all of nature is permeated by Divinity, but whether Mind, Consciousness, and Divinity even exist! Those who continue to rail against the evils of pantheism seem somehow to have missed the big picture.

If the caretaking of the soul has been thought in the past to be solely an exercise of the individual concerned about his personal fate, this may no longer be the case. Today it is the fate of the Earth, not just of single persons, that is also our concern. Can the recovery of the individual soul help in the rescue or preservation of our imperiled planet? Perhaps. The brilliant analytical psychologist James Hillman has reintroduced the concept of the world soul—the *anima mundi*. The world soul is not a remote thing but a power that permeates the world and all in it. Importantly, the soul of the world and that of the individual are inseparable, the one always implicating the other. As Hillman states, "Any alteration in the human psyche resonates with a change in the psyche of the world." If he is right, there is global significance in paying attention to our own soul. Indeed, the fate of the Earth could depend on our collective efforts.

If recovering our souls and becoming aware of the nonlocal nature of our own minds are such vital tasks, some readers may wonder why there are no specific paths laid down in what follows on "how to do it." The reason is that this is not a "how to" book. Ways of waking up to our inner divinity have been prescribed since the dawn of history and are contained within all the great spiritual traditions. Specific instructions in these methods are beyond the scope of this book.

If we continue in the ways in which we have conceptualized ourselves for hundreds of years, it is no longer certain that we will have a future on this Earth. If we are to survive, a sacred regard for the Earth and all things in it must arise once more. One way this can come about is through the essentially spiritual act of recovering our own soul—of waking up to our nonlocal Self, that immortal part of us that transcends space, time, and person. Without this awareness we will continue to act from isolation, fear, and defensiveness, and we shall continue to wreak disaster on all of nature and each other.

That is what this book is all about: the rediscovery of our soul-like, nonlocal nature—the re-realization that we are eternal, infinite, and One.

LARRY DOSSEY, M.D.
Dallas, Texas

Immortal . . .
Here, there, everywhere,
Extending to the farthest reaches of the cosmos . . .
Enveloping all minds and all persons . . .
Timeless . . . beyond death . . .
Existing at all moments,
Including all of the past and all of the future,
Forever and forever. . . .

This is a portrait of every person who has ever lived.
This book will show why it is true.

Part I

BEYOND THE BRAIN: The Evidence

The assumption of being an individual is our greatest limitation.

—PIR VILAYAT KHAN

1

The Reach of the Mind

The Mind's Eye

All suffering
prepares the soul
for vision.

—MARTIN BUBER

The surgery had gone smoothly until the late stages of the operation. Then something happened. As her physician was closing the incision, Sarah's heart stopped beating. Was it a reaction to the anesthetic? An undetected electrolyte disturbance in the blood? A result of some subclinical heart problem? The cardiac monitor suddenly showed ventricular fibrillation, a wild, chaotic, electrical storm in the heart in which no effective beat takes place. But the emergency was all over in a minute, for it took no more time than that for the anesthesiologist to defibrillate her with the LifePak device that was always at the ready in the OR, the operating room.

Yet Sarah had something else to show for her surgery experience besides the ache in her side where the stone-filled gall bladder had been removed and the concentric, reddish rings on her chest left by the sting of the defibrillator's paddles. She had something else to show that amazed her and the rest of the surgery team as well—a clear, detailed memory of the frantic conversation of the surgeons and nurses during her cardiac arrest; the OR layout; the scribbles on the surgery schedule board in the hall outside; the color of the sheets covering the operating table; the hairstyle of the head scrub nurse; the names of the surgeons in the doctors' lounge down the corridor who were waiting for her case to be concluded; and even the trivial fact that her anesthesiologist that day was wearing unmatched socks. All this she knew even though she had been fully anesthetized and unconscious during the surgery and the cardiac arrest.

But what made Sarah's vision even more momentous was the fact that, since birth, she had been blind.

Although her surgeon patronizingly dismissed her reports of the emergency when she questioned him, through her fog, on his first post-op visit to her in the intensive care unit, the critical care nurse was sympathetic: "It happens pretty often around here after anesthesia. People come back with the strangest stories. A couple of years ago one guy had a cardiac arrest, and when he came to he told his cardiologist what the specific levels of all his cardiac enzymes in his blood would be for the next three days. The doctor didn't believe him, either, but he hit it right on the money! Could have saved doing some lab tests if his doctor had taken him seriously!"

Eventually, although Sarah continued to turn the event over in her head endlessly, she reserved it for her own reflection; for, with the exception of a few of her nurses, no one wanted to discuss it. But it remained a stunning experience for her, for it clearly showed that, as she put it, "There's more than one way to see!'' Sarah now knew that the world worked differently than anyone supposed, that there were principles operating beyond the common view. No matter how poorly these ideas and her experience fit her previous model of reality, she could not dismiss them. She felt the impact of the event changing her. My vision cannot be completely in my body, she told herself, and it cannot really all be in my eyes and my brain. When my body was least functional during the arrest, my senses were most functional! The experience of her extraordinary perception continued to give her strength in her private world of blindness long after she left the hospital.

A Trip Inside the Body

> *What you see with your eyes shut*
> *is what counts.*
>
> —LAME DEER
> Sioux Medicine Man [1]

One bright Tuesday morning in early December, Elizabeth took a trip. She'd planned the journey for months, except for one important detail: she had intentionally never set a date for departure. Elizabeth wanted to let the trip happen on its own and knew she would understand when the time was right. This was the morning. Without ever leaving her house, Elizabeth traveled—into her body.

It was not an aimless excursion. Elizabeth wanted to know the reasons behind the nagging pain she had been experiencing for several months in the lower left side of her abdomen. Always given somewhat to introspection, she was confident that she *could* find out the reason behind it—moreover, that it was indeed her rightful task to do so. This job did not belong to others, even to her doctors, except perhaps as confirmation of her own knowledge. She did not know where her certainty came from, but it was there: an unpretentious, inner awareness that she could know her own body's ways.

She sat in her favorite wingback chair in a sunny spot by the big bay windows in the living room and closed her eyes. Her feet resting solidly on the floor, palms down on the chair arms, she took several slow, deep breaths and let her mind be free. This state of consciousness was not new to Elizabeth; she had regularly engaged in deliberate relaxation techniques for a decade and could enter deep states of tranquility at will. As usual she adopted no particular mental strategy, but let her consciousness be blank. I already know everything I need to know, she said to herself; I only have to let the knowledge surface. So, allowing her mind to be empty, she waited.

Images came: something multiple, circular, ovoid, now inside a larger object, itself seemingly spherical. Soft, white, three of them. Not angry, this triad, but calm and placid—something that *belonged* inside her and something that, strangely, seemed to own pain as a rightful expression of its being. Something unmistakably *her*.

For a half-hour the image changed little. Attaching no meaning to it, she let it be. Over this period of time she developed a certainty that the image had a life of its own and that she had not made it up through some process of psychological manufacture. She let the image recede, slowly.

A few days later she came to see me for a routine medical appointment.

"Look, I know I don't have cancer, but my gynecologist is really concerned. He says I have an enlarged ovary. He wanted to do more tests, but I told him I wanted to check it out first with my internist. Anyway, I already know what's wrong. And it's not cancer!"

Perfectly at ease, Elizabeth, an attractive, middle-aged woman, patiently described to me why she had come to my office. I understood her gynecologist's concern. An enlarged ovary in her age group had to be considered cancer until proven otherwise.

"How do you know it's not cancer?" I queried.

"I can *see* my ovaries. Both of them. Clear as day."

"What do they look like?"

"Right one's normal. But on the left one there are white spots. Three spots."

The room was silent as she waited to see if I, like her gynecologist, would reject her self-diagnosis. Something about this woman was extraordinary. She had a presence that commanded great respect. She was delivering what for her was a simple, honest statement of fact. There was nothing boastful about her ability to see these three white blemishes, no rancor toward her doctor for his disapproving attitude.

"Want to prove you're right?" I offered.

"How?"

"Let's do a pelvic examination and then a sonogram. Won't hurt a bit and it doesn't require any X rays. It'll give us a world of information and will tell us where to go from here. And if it's normal we'll both be happy."

She agreed. After the examination I led her down the hall to the sonography area and spoke to Jim, the radiologist on duty that day. While she was changing into an examination gown for the procedure, I took Jim aside and laid out the problem to him.

"Look, Jim, this woman has some pretty definite ideas about what's wrong with her. She maintains she has three little white spots on her left ovary. She has some chronic pain in the area and her gynecologist is worried about cancer—the left ovary's enlarged on pelvic exam. How about letting me know as soon as possible what you find on sono."

At the mention of the fact that Elizabeth "knew" her diagnosis, Jim smiled knowingly. He'd seen patients many times who "knew" what was wrong, only to be proved incorrect under the scrutinizing eye of some scientific test—an X ray, sonogram, or CAT scan. He wouldn't be critical of her belief.

He'd respect her views, he assured me with a pat on the shoulder, but this was followed by a look that said, "But we both know she's wrong!"

I went back to my office only to be interrupted a half-hour later in a conversation with a new patient by a loud knock on my door. Excusing myself, I rose to answer it. It was the radiologist, and he was obviously upset. He was gesticulating in uncharacteristic fashion and was speaking so fast I could hardly understand him. I stepped into the hall for privacy, closing my office door behind.

"Jim, not so fast. I can't understand a word you're saying. What's bothering you?"

"That woman's crazy!" he stammered. I'd never seen him so off balance.

"Never mind the psychiatric diagnosis, Jim. What did the sonogram show?"

Still angry, still fuming about his encounter with Elizabeth, Jim struggled to find the words.

"Three little white spots, just like she said! That's what she's got, all right, three cystic lesions which look precisely like small white areas, benign-appearing and all on the left ovary, the right ovary normal, and no sign of cancer anywhere! Dammit, how the hell did she know?"

"Well, Jim, with diagnostic skills like that, she could obviously put us both out of business."

Elizabeth's pain continued in the days that followed. I referred her for a second opinion to another gynecologist, a female physician cordial to her views. They developed a warm relationship. Eventually Elizabeth grew weary of the pain, and both she and her gynecologist agreed that surgery seemed indicated. The operation, like the sonogram, confirmed Elizabeth's diagnostic acumen: on the left ovary were three completely benign cysts. Following surgery, which was uneventful, her pain disappeared completely and she remains well.

The radiologist's and Sarah's doctor's reactions of automatic disbelief rejection are typical of many people who encounter the enormous capacities of the mind for the first time in medical situations. Everyone "knows" that the mind can only do so much. It is lodged in the brain, a passive receptor and processor of sensory data, dependent on the brain's anatomy and physiology, a by-product of biochemical reactions that occur in brain tissue which, if destroyed, causes the destruction of the mind as well.

But these prejudices about how the mind should function don't fit with the clinical reality of how it *does* function, as an enormous body of evidence in medicine attests. Apparently the mind never got the message about how it ought to behave.

If cases like Elizabeth's were rare, perhaps they would deserve no more than a raised eyebrow. But they aren't. Medicine is thickly littered with similar examples showing that the mind's range is beyond the brain.

Frequently it appears that an illness or a crisis with one's health is the key that turns on nonlocal ways of knowing, as in Sarah's and Elizabeth's cases, freeing the mind to behave nonlocally in space and time. Let's continue our search for more facts that don't fit our traditional models of the mind.

Telling the Body What to Do

> *I can no longer make a strong distinction between the brain and the body. [2] . . . the research findings . . . indicate that we need to start thinking about how consciousness can be projected into various parts of the body. [3]*
>
> —CANDACE B. PERT
> Former Chief of Brain Chemistry
> National Institute of Mental Health

> The subject is a 39-year-old woman who has followed an Eastern religious practice for the last nine years. During most of this time, as part of her religious practice, she would usually meditate once or twice daily for about 30 minutes. For the last three years, she has followed a specific tantric generation meditation practice whereby "higher energies" are visualized and she seeks to transform herself into those energies.
>
> During [this experiment] she would usually reserve about five minutes of her daily meditation for attention to [this] study. First she would dedicate her intention concerning the study for universal good instead of self-advancement. She would also tell her body not to violate its wisdom concerning her defense against infection. Finally she would visualize the area of erythema and induration [the reddish, hard area surrounding the skin test for the infectious disease] getting smaller and smaller. . . . she would pass her hand over her arm [the site of the skin test], sending "healing energy" to the injection site. [4]

These descriptions come from the report of perhaps the first fully documented case of a human being intentionally changing the function of the immune system. This experiment and many others like it bring the power of science into the task of showing that the mind can extend its influence far beyond the brain. It shows that the mind can penetrate to the cellular level of the body and modify "mindless" bodily processes. No longer in the category of mere folk wisdom or superstition, the body-mind connection is now a matter that has been demonstrated by careful scientific inquiry.

The case above was reported in a prestigious medical journal by a team of investigators at the University of Arkansas College of Medicine led by Dr. G. Richard Smith. As with most important scientific studies, it rested on a solid background of prior observations by other workers in the same field. For instance, in 1963 Stephen Black and his British colleagues had found that hypnosis could be used to inhibit the body's immune response. They suggested to a group of subjects under hypnosis, who had positive skin test reactions to tuberculosis, that they would not react to further skin testing for the disease. A positive skin test means that one either has been exposed to the disease in the past or has an active infection. Once a test is positive, it remains that way for life. In spite of the fact that their skin tests were strongly positive, on retesting the subjects markedly inhibited the degree of their reaction to the injected skin test material. [5] This was a decidedly odd finding, for it is universally believed that reactions to skin tests for infectious diseases such as tuberculosis are a purely "body" process, completely uninfluenced by the psyche. But Black's findings showed otherwise.

The result of this experiment encouraged Smith and his associates to pursue the effects of the mind on the body's immune reactions. They selected seven persons who had strongly positive skin tests to tuberculosis. These subjects were then exposed to a behavioral conditioning sequence in which they were trained to expect a positive reaction on the arm injected with the actual tuberculosis skin test material, and to expect a negative reaction on the other arm, which was injected with normal saline (salt water) as a control. Then, unknown to them, the saline and the tuberculosis skin test substance were switched. When this was done their bodies continued to react according to their *expectations,* not the actuality. Their reactions to the tuberculosis skin test material, which they now believed to be saline, were reduced from an area of redness and hardness with a mean of 15 millimeters diameter to one of only 4 millimeters diameter. Their expectations seemed capable of overriding the purely automatic physical responses of the body. Their bodies had clearly begun to behave according to the dictates of their thoughts and not according to mindless internal processes. [6] But all these experiments occurred without the full awareness of the subjects. They had either been hypnotized or behaviorally conditioned against their knowledge. What would happen, Smith and his colleagues wondered, if someone tried to *consciously and intentionally* modify his or her own immune system? Leaving hypnosis and behavioral conditioning behind, they now chose to work with a subject who was skilled in directing the effects of the mind, the thirty-nine-year-old woman in the above citation.

What was also astonishing, in addition to the fact that she could markedly inhibit her positive reaction to the viral skin test material, was that the inhibition of her immune cells, called lymphocytes, persisted even when they were removed from her blood and studied outside her body. As the scientists stated, "It appears that the subject, acting with intention, was able to affect not only her skin test response but also the response of her lymphocytes studied [in test tubes]." Thus Smith and his colleagues, in the dry jargon of scientific reportage, described what may be "the first published data of an intentional direct psychological modulation of the human immune system." [7]

Interpreting the Results: Is It Mind or Chemistry?

What should we make of this case report and the many similar ones that have accumulated since it was published? Most researchers believe there really is nothing extraordinary going on here, that all is explainable in the sophisticated language of neurotransmitters, neurons, synapses, and receptor sites. "Higher central nervous system events"—physiological jargon for thoughts—trigger the release of hormones into the blood, or they stimulate electrical nerve signals that travel from the brain, down the spinal cord, and into other areas of the body. These events cause the body to do what it does. Through these neural and hormonal processes we can explain how a skilled meditator could inhibit her immune system's response to foreign material that has been injected into the skin. This explanation regards the mind much as bioscience has always seen it: the manifestation of extremely complex events originating in the brain, the effects of which are translated throughout the body. Essentially similar processes are going on whether we are moving a finger, opening a car door, or inhibiting our immune response. And the failure of the subject's lymphocytes to react to stimuli outside her body in the test tube? Again, nothing mysterious: undoubtedly they had been affected in some way before the blood was removed from her body, explainable again by purely physical mechanisms.

The subject's own explanations are not scientific, not reliable, say many scientists. There are really no "healing energies" at work and she is not really "telling" her immune system to react in a certain way. If any messages are being delivered they are not mental but physical, in the form of nerve impulses and hormonal events. One need not invoke "Tantric higher energies" to explain what happened in this case study. In fact, one need not introduce "mind" at all, for even the mind can be explained by the chemical processes going on in the brain.

But what if the mind is nonlocal? What if it is not *in* space? Then the traditional neural and chemical explanations are inappropriate, for they force characteristics onto the mind that are not applicable—attributes such as locality, of confinement to the physical brain. This is not to say that the brain's chemical processes do not exist or that they are not important. Indeed, today, no one would deny the existence of a profound *relationship* between the brain's chemistry and our mental processes. But the fact that such a relationship exists may not mean that the mind is completely confinable to the brain.

But what *are* these chemical events within the brain if not the equivalent of the mind? After all, we are confronted with evidence for this "equivalence" every day in medicine. In a stroke, for example, damage to the brain may cause profound impairment of a patient's mental abilities. Surely this must mean that the mind must be inside the brain. But that is by no means a necessary conclusion. As an example frequently used by biologist Rupert Sheldrake to show that memory may not be confined to the brain, think of your TV set. A search inside your TV for traces of programs you watched last week would be doomed to failure, for the simple reason that your set tunes in to

TV transmissions but does not store them. Moreover, damage to the set might destroy certain channels or diminish the clarity of the reception. Yet this is not evidence that the sounds and pictures you see originate inside the TV set itself. There is thus the closest *relationship* or *correlation* between the inner workings of the TV set and the television signals. But "relationship" and "correlation" are not the same thing as "equivalence." We must be careful, then, in jumping to the conclusion that since the brain is entirely local in space and time, so too is the mind.

Where Is Love?

Everyone recognizes the existence of nonmaterial phenomena that are quite real but that do not have specific lifetimes or locations. Where, for example, is love? Or patriotism? Or loathing of the taxman? Where is history? All these emotions or concepts are not really things. They do not occupy points in space, nor do they occupy spans of time in the same sense that a bird or a mouse or a tree have a demonstrable lifetime. Yet they are real. And they obviously "do" things in the world in spite of their nonlocality in time and space and in spite of their nonmateriality. Love and patriotism are some of the most potent forces known to humanity, yet they have never been seen, measured, or described by differential equations. We do not expect locations of them. Assigning them coordinates in space and time seems silly. We do not suppose that we shall ever find them three blocks from home or four feet above our heads. Neither do we expect them to appear at half past twelve or in A.D. 1087, nor to endure for twenty minutes or three days. Just so, many powerful forces in our lives declare themselves in the most vivid fashion while being completely invisible, nonlocal, and nonmaterial.

In his important book, *God and the New Physics,* physicist Paul Davies discusses the error of regarding certain concepts as things requiring a location and made of some sort of stuff. "What stuff is the soul made of?" he asks. "The question is as meaningless as asking what stuff citizenship or Wednesdays are made of. The soul is a holistic concept. It is not made of stuff at all. Where is the soul located? Nowhere. To talk of the soul as being in a place is as misconceived as trying to locate the number seven, or Beethoven's Fifth Symphony. Such concepts are not in space at all.'' [8]

What, then, is odd in thinking about the mind nonlocally? Usually the answer is that the mind *seems* so obviously part of the brain, so much a part of our body. But in fact, our predecessors would have thought this a rather strange thing to believe. As psychologist Lawrence LeShan points out, ''It was not until Alcmaeon studied the relationship of brain injury and consciousness in the sixth century B.C. that the relationship of consciousness to the brain was known by Western society, and there are still a good many 'primitive' cultures that locate consciousness in other areas of the body, if they locate it anywhere at all.'' [9]

But the fact that the mind may be nonlocal does not mean that it could not act *through* the brain. The concept of ''Sunday'' may act through a person, motivating him to attend church or picnic in the park. Every day we see countless nonmaterial concepts manifesting themselves in quite local and physical ways. Some, such as courage, valor or fear, have demonstrable physiological correlates within the body (for example, adrenalin) that are well known. In the same way, a nonlocal, nonmaterial mind might manifest its existence via the brain, creating the cascade of neurohumoral events that the physiologists describe in minute detail.

Descartes was perhaps the most prominent Western philosopher who was troubled about this matter. As a solution to how nonmaterial mind could affect the material brain, he postulated that the pineal gland, deep within the cranium, was the mysterious site where this murky transaction could take place. But his suggestions were never convincing. Some of Descartes' disciples tried to meet the objections by inventing fantastic models of parallel worlds that ran in synchrony but never interacted. The material and the mental worlds could therefore exist side by side without ever overlapping. This way the problem was bypassed but again unsatisfyingly, and today there are few people who believe in parallel worlds (although it must be said that the idea has surfaced again in modern science as a response to the so-called observer problem in quantum mechanics).

Today we have simply gotten used to the philosophy that something nonmaterial cannot possibly affect something material. Yet this assumption should be put in its historical perspective, as philosopher and parapsychology researcher John Beloff has done. Beloff describes the way this idea emerged:

> In fact, . . . Descartes and his followers had simply taken over a principle of scholastic philosophy according to which an effect must be of the same nature as its cause. But there is no reason in logic why this must be so. There is nothing contradictory in supposing that an immaterial entity, if that is what the mind is, could produce physical effects. . . . The idea still being purveyed by so many modern philosophers, that there is something absurd or incoherent about [this], . . . is without foundation. [It is] quite right to insist that mind and matter must have "something in common" but that something is, precisely, the power to influence one another; nothing more than that is required. [10]

Today serious thought is being turned once more to the possibility that nonmaterial entities may affect material things. One of the most rigorous proposals of this sort has been put foward by Professor Henry Margenau of Yale University, who for many years has made fundamental contributions to the theories underlying quantum physics. In Part II we will examine his suggestions in greater detail. But the point we want to propose now is that local, physical manifestations do not necessarily require local, physical causes. It would be as foolish to say that the local brain demands a local mind as to say that the individual love poem requires that love be confinable, or that the raised flag shows that patriotism can be nailed down in space and time.

Unity Consciousness: An Everyday Experience

Must one be unhealthy or near death for consciousness to manifest beyond the brain or body, as in the accounts of Sarah and Elizabeth? The answer seems to be no: Beyond-the-body experiences, while common during illness and near-death situations, are apparently distributed ubiquitously in the everyday lives of quite ordinary people.

J.B. Priestley, the English playwright and man of letters, had a lifelong fascination with the nature of reality. His book *Man and Time* is one of the best incisive modern commentaries on the nonlocal nature of the world. Priestley was particularly fascinated with precognitive dreams, in which people can know of an event before it actually occurs. This is a genuinely nonlocal event in time in which the mind violates the barrier between the present and the future. Priestley broadcast an appeal over the BBC for people to send him anecdotes of paranormal experiences. He was deluged with mail. Priestley analyzed this outpouring scrupulously. Even after allowing for instances of self-delusion, he still had a solid body of evidence that left him with an unequivocal conclusion: The ability to know reality via precognition, telepathy, or clairvoyance was a common human trait. Commenting on the nonlocal Self as it manifests in dreams, he concluded that,

> There is nothing supernormal and miraculous about this larger temporal freedom of the dreaming self. It is not a privilege enjoyed by a few very strange and special people: It is part of our common human lot. We are not—even though we might prefer to be—the slaves of chronological time. We are, in this respect, more elaborate, more powerful, perhaps nobler creatures than we have lately taken ourselves to be. [11]

In contrast to Priestley's view is the common belief that persons who have nonlocal mental experiences are somehow deranged or off balance psychologically. The minds of "normal" people don't go wandering off; they stay put in the body and the here-and-now. Sociologist Andrew Greeley, of the University of Chicago's National Opinion Research Center, disagrees. He reports that people who have transtemporal and transspatial experiences are "anything but religious nuts or psychiatric cases. They are, for the most part, ordinary Americans, somewhat *above* the norm in education and intelligence and somewhat *less* than average in religious involvement." Among the people Greeley tested were persons who had had profoundly mystical experiences, such as being bathed in white light. When they were subjected to standard tests that measure psychological well-being, they scored *at the top*. University of Chicago psychologist Norman Bradburn, who developed one of the tests, said that no other factor had ever been found to correlate so highly with psychological balance as did mystical experience. [12]

The Suddenness of Nonlocal Experiences

One of the typical qualities of nonlocal experiences is the suddenness with which they appear. Take the case of Rosalyn Tureck, the renowned concert artist who was the first woman invited to conduct the New York Philharmonic Orchestra, and the author of several books, including one that links the structure of Bach's music to two physical theories. Shortly before her seventeenth birthday she was playing the Bach fugue in A minor from Book 1 of the Well-Tempered Clavier. Suddenly she lost all awareness of her own existence. On coming to, she saw Bach's music in a totally new way, with a new structure that required the development of a novel piano technique. She worked it out over the next two days, applying it to four lines of the fugue, which she played at her next session. Her teacher felt her interpretation was marvelous but could not be sustained or applied to Bach's entire oeuvre. "All I knew," Tureck said, "was that I had gone through a small door into an immense living, green universe, and the impossibility for me lay in returning through that door to the world I had known." [13]

Beyond Language

How do creative persons actually navigate in that "immense living, green universe"? There is an enormous body of evidence suggesting that the verbal, linear, sequential, logical mental processes that make sense in ordinary waking life are not involved. In fact, in the domain of experience whence creativity erupts, language itself seems to play little or no part. As Arthur Koestler stated in his monumental treatise on creativity, *The Act of Creation,* "Language can become a screen which stands between the thinker and reality. This is the reason why true creativity often starts where language ends." [14]

The evidence suggests that even for creative scientists, physicists, and mathematicians, language is overrated and is not essential to creativity. In 1945 the mathematician Jacques Hadamard conducted a survey of the most eminent mathematicians in America. He wanted to find out about their working methods. In response to his questionnaire, Albert Einstein replied,

> The words or the language, as they are written or spoken, do not seem to play any role in my mechanism of thought. . . . Conventional words or other signs have to be sought for laboriously only in a secondary stage. . . . when words intervene at all, they are, in my case, purely auditive. [15]

Einstein's case was by no means unusual. Hadamard concluded from his 1945 survey that practically all the mathematicians born or resident in America avoid not only the use of "mental words" but even "the mental use of algebraic or other precise signs. . . . The mental pictures [that they employ] are most frequently visual." [16] This conclusion hardly fits the cartoonlike images many persons have of an infinitely large cloud of differential equations constantly floating above the mathematician's head.

Not only does language seem to be bypassed in the creative process, some of the physicists who have demonstrated the highest levels of creativity have apparently done so without benefit of mathematical thinking at all. Perhaps the most astonishing case is that of the English physicist Michael Faraday, whom Einstein placed on a par with Newton. Faraday's thinking was almost entirely visual, and strikingly devoid of mathematics. Indeed, he had neither a mathematical gift nor any formal training in mathematics, and he was ignorant of all but the simplest elements of arithmetic. Yet Faraday could "see" the stresses surrounding magnets and electric currents as curves in space, and he coined the phrase "lines of force" to describe them. In his mind they were as real as if constituted of solid matter. Faraday saw the entire universe made up of these lines of force, and he saw light as electromagnetic radiation. His visualizations led not just to theories but to practical results, including the invention of the dynamo and the electric motor. In 1881, Hermann von Helmholtz, one of the greatest mathematical physicists of his day, paid homage to Faraday's remarkable ability in his Faraday memorial lecture:

> It is in the highest degree astonishing to see what a large number of general theorems, the methodical deduction of which requires the highest powers of mathematical analysis, he found by a kind of intuition, with the security of instinct, without the help of a single mathematical formula. [17]

Not only language and mathematical thinking but rational thought in general may be overvalued in the creative process. On the basis of his historical survey of a large number of creative geniuses in science, Koestler came to the conclusion that "their virtual unanimous emphasis on spontaneous intuitions, unconscious guidance, and sudden leaps of imagination which they are at a loss to explain, suggests that the role of strictly rational thought processes in scientific discovery has been vastly overestimated since the Age of Enlightenment; and that, contrary to the Cartesian bias in our beliefs, 'full consciousness,' in the words of Einstein, 'is a limit case.' " [18]

The greatest creators do not think or reason so much as they simply "see." Perhaps, then, it is not coincidental that we have always referred to our greatest scientists, writers, and painters as "visionaries" and "seers."

The Wholeness of Nonlocal Experiences

Besides being sudden, nonlocal experiences also manifest a typical quality of wholeness. This is particularly evident in dreaming and reverie, wherein complex revelations can emerge fully formed. This suggests that they have no history, that they were not laboriously worked through, bit by bit, in time. Mozart is a famous case in point. As he described his experience with composition,

> All this fires my soul, and, provided I am not disturbed, my subject enlarges itself, becomes methodized and defined, and the whole, though it be long, stands almost complete and finished in my mind, so that I can survey it, like a fine picture or a beautiful statue, at a glance. Nor do I hear in my imagination the parts *successively,* but I hear them, as it were, all at once. What a delight this is I cannot tell! All this inventing, this producing, takes place in a pleasing, lively dream. [19]

Intuition: Beyond Space and Time

The inventor of the electron microscope, Max Knoll, who for many years was a professor in the Department of Electrical Engineering at Princeton University, was intrigued by both the suddenness and wholeness with which new ideas frequently appear. He linked this process to intuition, which he viewed as a genuinely nonlocal phenomenon that breaks the constraints of space and time:

> The fact that an idea suddenly emerges full-blown call[s] for the existence of . . . a special intuitive function. The content of this idea is best described in . . . timeless, nonspatial . . . terms. . . . Always unmistakable are the suddenness and activity of the intuitive event, and its tendency to occur in a state of relaxation, and after a protracted 'period of meditation.' . . . This . . . cannot be attributed . . . to higher thinking functions. [20]

No one knows where intuition comes from. The word itself is derived from the Latin *intueri,* meaning "to consider, to look on," suggesting some outside vantage point. Several candidate theories have been offered through the centuries. One proposal is that there is a realm of preexistent ideas, such as the mind of God or the realm of Ideas that Plato described, where this nonsensory information might originate. Somehow we tap into this domain and later call the entire process "intuition," by which we "just know" something. Alternatively, the psychiatrist C.G. Jung suggested the existence of inherited primal forms of thought, "archetypes," which serve as an everpresent reservoir of knowledge that we might tap without knowing it.

Most recently, in contrast to these philosophically and psychologically based theories, scientists have got into the act of accounting for intuition. Based on the results of so-called split brain research, neurophysiologists such as Nobelist Roger Sperry have proposed that the intuitive faculty may have a neurological explanation, residing in the right cerebral hemisphere of the brain in most persons. Intuition is explained as a nonverbal, nonlogical process that is difficult to communicate in words, in contrast with the linear, discursive, logical/rational mode of thinking that takes place primarily in the left side of the brain.

Sometimes, Jung stated, when we exercise our intuition we experience a deeply satisfying sense of having done things "right." Could this be due to an immersion in the True Self, the One Mind, in which we see the world as it really is—timeless, unbounded, and whole?

Sometimes the intuition of different persons seems to operate in synchrony, leading to simultaneous discoveries. One of the most famous examples of this phenomenon in the history of science is the codiscovery by Wallace and Darwin of the theory of evolution. Could simultaneous discovery be due to the absorption of individual minds in the Universal Mind, whereby information could conceivably be shared by everyone?

Nonlocal Experiences: Why Not More *Common?*

If our minds can manifest nonlocally in all the ways we have mentioned—dreams, visions, intuition, and so on—why are we not more aware of these types of experiences? Why are they not more common? Philosopher William Irwin Thompson suggests the reason is that we are suffering a kind of collective hypnosis, a cultural trance, that prevents us from seeing things the way they really are. "We are like flies crawling across the ceiling of the Sistine Chapel," he states. "We cannot see what angels and gods lie underneath the threshold of our perceptions. We do not live in reality; we live in our paradigms, our habituated perceptions, our illusions; the illusions we share through culture we call reality, but the true . . . reality of our condition is invisible to us." [21]

A different answer was given by the philosopher Henri Bergson, who pointed out that we distort our sense of reality by the language we use, especially when we speak of space and time. Bergson said, "If we want to reflect on time, it is space that responds"—as when we say the past lies *behind* us and the future somewhere *ahead* of us. [22] We thus make time a prisoner of space by always expressing duration as an extension. This habit causes great difficulty in comprehending what nonlocality is all about. When we say, for instance, that eternity—a nonlocal concept—is "a long time," we imagine that it is lots of serial moments strung together; and when we reflect on infinity—another nonlocal concept—we suppose it, too, is lots of spatial points arranged in linear sequence. But eternity and infinity are not bit-by-bit phenomena; they are wholes.

Why do we spatialize time? The renowned physicist Bernard d'Espagnat, who has written widely about nonlocal reality in modern physics, observes that human beings feel much more at home in a world of solid things than of fluid or invisible things. Solids have shapes and forms. They are extended in space and are confined to specific places: They are *local*. Could it be that our preference for the local is therefore rooted in our biology, explaining some of our difficulty in noticing nonlocal experiences when they do occur? In his book, *In Search of Reality*, d'Espagnat states,

> Human intelligence . . . is above all a knowledge of solids: this is a result of the evolution of our species, whose struggle for life was specifically based on armaments and on tools made of solids, in contrast to the animals. Consequently . . . [we are] incapable of capturing life, to the extent that the essential thing in the latter is the fluid, the continuous, the moving. [23]

Added to the biological factors are important cultural elements that influence how we notice the nonlocal facets of reality. No one can doubt that we have made a fetish of time, becoming what Jeremy Rifkin calls "the nanosecond culture." [24] Adherence to locality in time—always looking ahead or behind, never dwelling in the moment—has become a morbid obsession. It has spawned a variety of physical illnesses that we could collectively call "time sicknesses"—coronary artery disease, hypertension, peptic ulcer disease, irritable bowel syndrome, and the vascular headache syndromes including migraine, to name only a few. In general, any illness in which anxiety and excessive time awareness have been shown to play a role belong in this growing category of human maladies.

Time awareness is a movement away from a nonlocal sense of time to a contracted, local awareness. This can be easily shown. In the Biofeedback Department at Dallas Diagnostic Association, where all of the above disorders are treated through biofeedback training, when a patient comes for his first visit we ask him to perform a very simple task. He is requested to lean back in the reclining chair, relax, close his eyes, and tell us when a minute has passed. *Everyone* underestimates. Our record holder for many years was a man suffering from extreme anxiety who blew the whistle at fifteen seconds; but this record was shattered by a gentleman with coronary artery disease whose estimate ended at twelve seconds! With biofeedback training the time sense invariably expands and elongates. On their final visit our patients are asked once again to perform this task. They regularly overshoot the mark, sometimes allowing up to three minutes to pass before signaling the passage of a minute.

Retraining the time sense in biofeedback, meditation, or other forms of deep relaxation increases the awareness of nonlocal happenings in one's life. Paranormal or extrasensory events regularly become more frequent; precognitive dreams may occur; intuition and creativity may flower. Thus, even if our difficulty in perceiving nonlocal manifestations is built into our biology and our culture, the situation is not hopeless, since our bodily senses can be retrained.

Jung: A Psychologist for the Everyday, One Mind

Frequently, when persons allow or train themselves to become sensitive to the many manifestations of nonlocal reality in their everyday experience, they become distressed on discovering there is no place to turn with this new knowledge. They may regard themselves as different and strange; they may feel isolated and alone; they may even come to see themselves as pathologically unbalanced.

Currently it is true that scientific medicine and psychiatry have little place for these everyday, nonlocal experiences. Today the definition of the "mental norm" is constricted; psychic experiences in which traditional concepts of space, time, and person are violated are generally dismissed or scoffed at. It is reassuring to realize, however, that within twentieth-century psychiatry there *is* a concept of mental functioning that is fully capable of absorbing nonlocal experience as healthy and normal: the model developed by the Swiss psychiatrist Carl Jung. Jung's own life was a profound example of how nonlocal experiences can erupt in everyday existence. From youth to old age he experienced frequent events that deeply influenced him in his development of a psychology of the transpersonal, the transspatial, and the transtemporal.

Jung is one of the greatest explorers of the mental life of human beings in our century. At first a collaborator with Freud, Jung parted with him because he believed Freud's theories stopped short of a full explanation of mind. Freud emphasized the repression of sexual instincts as an explanation for the afflictions of man's psyche—an overwhelmingly "local" model of mental illness that emphasized rigid categories of time, space, and person. Jung went further, developing a thoroughly nonlocal model of the psyche.

Many of Jung's radical insights resulted from his own experiences, particularly his dreams. "Day after day we live far beyond the bounds of our consciousness," he concluded. "Without our knowledge, the life of the unconscious is also going on within us . . . communicating things to us . . . synchronistic phenomena, premonitions, and dreams." [25]

As a result of analyzing his own psychic life across decades, as well as treating thousands of patients and analyzing their dreams, Jung became convinced that mankind possessed a definite *psychic heredity*. This consists of phenomena essential to life and which express themselves psychically, just as other inherited characteristics express themselves physically. Among these are "psychic factors" that are not confined either to single persons, families, or races. These "universal dispositions of the mind" are analogous to Plato's forms or to logical categories that are everywhere present as basic postulates of reason—the difference being that they are categories of the *imagination,* not categories of reason. Following St. Augustine, Jung called them *archetypes*. They abound in the lives of everyone and take the form of familiar motifs—religious stories, myths, dreams, spontaneous fantasies, and visions. The unconscious layer of the psyche that is made up of these universal dynamic forms Jung called the *collective unconscious*. [26]

Jung found that the collective unconscious demonstrates the traits of nonlocal mind we have seen so far. It could not be pinned down in space or time, and it transcended the single self to envelope all minds. As he put it, "The unconscious . . . has its 'own time' inasmuch as past, present, and future are blended together in it." [27]

There can be no doubt that Jung also believed in the concept of the One Mind, a theme we will develop later. "Since all distinctions vanish in the unconscious condition," he said, "it is only logical that the distinction between separate minds should disappear too. Wherever there is a lowering of the conscious level we come across instances of unconscious identity." [28]

Jung was aware that one of the common manifestations of the unconscious mind was the basic mystical sense of oneness and unity with all there is. This experience, like the collective unconscious itself, was universal. Jung called it "the transcendent at-one-ment" that placed one in contact with the One Mind. But ultimately this universal Mind and the single mind were one and the same. "Does [this] mean that the Mind is 'nothing but' our mind? Or that our mind is the Mind? Assuredly it is the latter . . . there is no hybris in this; on the contrary, it is a perfectly accepted truth [in the East], whereas with us [in the West] it would amount to saying 'I am God.' "

But although it may seem blasphemous to the Westerner to acknowledge such a thing, this realization, Jung stated, was nonetheless an "incontestable 'mystical' experience" present in all religious traditions, East *and* West. [29]

Jung firmly believed in immortality, which fit with his belief that Mind is beyond the limitations of time. "[Our] psyche reaches into a region held captive neither by change in time nor by limitation of place," he said. "The two elements of time and space, indispensable for change, are relatively without importance for the psyche. . . . [The] psyche is up to a certain point not subject to corruptibility." But the situation is paradoxical: In order to know immortality we must equally realize that we are mortal. "This feeling for the infinite," he maintained, ". . . can be attained only if we are bounded to the utmost. In knowing ourselves to be unique . . . that is, ultimately limited . . . we possess also the capacity for becoming conscious of the infinite. But only then!" [30]

Paying attention to the manifestations of the timeless Mind was for Jung the redeeming life task for all persons. This task is especially difficult in our era because we have shifted all emphasis to the here and now—to doing, to consuming, to practicality, to material "progress." But the One Mind cannot be put in a box in the here-and-now because it is infinite and eternal. And because its "brand" of space and time is different from that which we commonly value we find ourselves cut off from it. The result is pathological: We have become a victim of our own unconscious drives, and a "daemonization" of our world has come about. On the contrary, our task in life, Jung asserted, is "the exact opposite: to become conscious of the contents that press upward from the unconscious"—"to create," as he put it, "more and more consciousness." Only in this way can we realize "the sole purpose of human existence[:] . . . to kindle a light in the darkness of mere being." [31]

Jung did not agree with the tendency of the Western religions to regard the soul as something pitifully small, unworthy, personal, and subjective, and he pointed out the inner contradictions of this view. For instance, how could something so small and unworthy be immortal, as these religions insist? There are clues all around us, he said, to the contrary—that the soul is magnificently unbounded—and he spent a lifetime accumulating evidence for this. His conclusion: "The soul is assuredly not small, but the radiant Godhead itself." [32]

The connection between the soul and consciousness? Jung felt that human consciousness is "the invisible, intangible manifestation of the soul." [33] Thus the task of "creating more and more consciousness" becomes the equivalent of recovering the soul and regaining contact with the inner Divinity. But how can this be done? Today this task is enormously difficult, mainly for two reasons. First, the declarations of science prohibit this realization, for science tells us that "souls" do not exist and that consciousness is only the result of the physical processes in the brain. Second, the religions in the West generally consider the idea that the soul is "the radiant Godhead itself" as either very dangerous or downright blasphemous, a position we will later examine. Thus in our culture, our two great vectors of knowledge and wisdom—science and religion—tell us that "creating consciousness" and "recovering the soul" are ill-conceived.

But what, then, will we do with the recurring evidence such as we have examined above that suggests there is something of us that is unconfined to bodies or brains, and, as we shall later see, unrestricted by space and time? The usual response has been to ignore it. But if we wish to go through these data and not around them, we must suppose that science and religion have not had the final word on the nature of consciousness and its connection to the soul. Things are simply more complicated and grander than we have been told.

Now we look at more evidence that we are "assuredly not small," beginning with the power of prayer.

2

The Power of Nonlocal Mind

What is here is also there; what is there,
Also here. Who sees multiplicity
But not the indivisible Self in all
Wanders on and on from death to death.

—KATHA UPANISHAD

Prayer: Old Approach, New Wonders

. . . anywhere
is the center
of the world.

—BLACK ELK
Sioux medicine man [1]

We now go to an area of healing that is off the beaten path but one that is making a comeback of late: the domain of prayer.

Prayer has long been held by most religious traditions to contain potent healing power and has a long and honored history as a form of intervention in illness. Yet it has never been recognized in orthodox medicine as anything more than superstition—something that can't hurt, but that can't help much, either. But "prayer researchers" have surfaced in medicine today. These scientist-clinicians offer us some of the most remarkable evidence that the mind may indeed be nonlocal and can act on matter in decisive ways—ways that may make the difference between life and death for the sick person.

Prayer and the Coronary Care Unit: Evidence for Nonlocal Mind

Most people today believe that in science there is no place for prayer. Perhaps this idea is a holdover from some three centuries ago, when "action at a distance" was deplored by the best minds. Galileo condemned Johannes Kepler's views on gravity as "the ravings of a madman" when the latter proposed that invisible forces from the moon, acting across gigantic distances, were causing the earth's tides.

Obviously the modern mind has sided with Kepler by accepting the action at a distance that is gravitation, but we have not been so generous in our attitude toward prayer. However, in perhaps the most rigidly controlled scientific study ever done on the effects of prayer, cardiologist Randolph Byrd, formerly a University of California professor, has shown that prayer works and that it can be a powerful force in healing.

Byrd designed his study as "a scientific evaluation of what God is doing." "After much prayer," he states, "the idea of what to do came to me." [2] During his ten-month study a computer assigned 393 patients admitted to the coronary care unit at San Francisco General Hospital either to a group that was prayed for by home prayer groups (192 patients) or to a group that was not remembered in prayer (201 patients). The study was designed according to the most rigid criteria that can be used in clinical studies in medicine, meaning that it was a randomized, prospective, double blind experiment in which neither the patients, nurses, nor doctors knew which group the patients were in. He recruited Roman Catholic and Protestant groups around the country to pray for members of the designated group. The prayer groups were given the names of their patients, something of their condition, and were asked to pray each day but were given no instructions on how to pray. "Each person prayed for many different patients, but each patient in the experiment had between five and seven people praying for him or her," Byrd explained. [3]

The results were striking. The prayed-for patients differed from the others remarkably in several areas:

1. They were five times less likely than the unremembered group to require antibiotics (three patients compared to sixteen patients).

2. They were three times less likely to develop pulmonary edema, a condition in which the lungs fill with fluid as a consequence of the failure of the heart to pump properly (six compared to eighteen patients).

3. None of the prayed-for group required endotracheal intubation, in which an artificial airway is inserted in the throat and attached to a mechanical ventilator, while twelve in the unremembered group required mechanical ventilatory support.

4. Fewer patients in the prayed-for group died (although the difference in this area was not statistically significant).

If the technique being studied had been a new drug or a surgical procedure instead of prayer, it would almost certainly have been heralded as some sort of "breakthrough." Even so, anyone can appreciate the striking implications of this study. Even hardboiled skeptics seem to agree on the significance of Byrd's findings. Dr. William Nolan, who has written a book debunking faith healing, acknowledged, "It sounds like this study will stand up to scrutiny. . . . maybe we doctors ought to be writing on our order sheets, 'Pray three times a day.' If it works, it works." [4]

What lessons does this study hold for us in our search to understand the mind and its role in medicine? The implications are far-reaching and take us into spiritual considerations that we will approach in later chapters. But for now we can emphasize some of the nonspiritual aspects of prayer that shed important light on the nature of the mind. This rigorous study suggests that something about the mind allows it to intervene in the course of *distant* happenings, such as the clinical course of patients in a coronary care unit hundreds or thousands of miles away. In this prayer study the degree of spatial separation did not seem to matter: Byrd did not discover that prayer groups just around the corner from the hospital were more effective than those hundreds of miles away.

This suggests immediately that there is no "energy" involved in prayer as we understand this term in modern science. The strength of energetic forces—for example, the intensity of a radio signal or a beam of light—rapidly decreases with increasing distance. If some sort of mental energy were involved in prayer, then praying persons in San Francisco, where the hospital is located, would be able to achieve better results than equally skilled persons in New York or Miami, since the energy would have less distance to travel and would be stronger when it arrived. But no such correlations were found.

This suggests that the effects of prayer do not behave like common forms of energy; that no "signal" is involved when the mind communicates with another mind or a body at a distance. As a result, when the mind works—here via prayer—distance is not a factor.

And if it is not, it is improper to think of the mind or our prayers as "going" anywhere. In spite of the fact that many who pray think that they are "sending" their prayers across space to the sick person, or that they are bouncing them off the Almighty back down to the sick person (God as communications satellite), we must not take these ways of thinking seriously because they do not fit with our observations. That is not to say that God is not involved, only that distance is once again not a factor. And on this fact, all the major theistic religions agree: They have never confined God to a specific place. He is everywhere, He transcends spatial confinement and location; He is nonlocal, an attribute shared by our own minds. Thus we can say without hesitation that something about us is divine.

The word "local," in common parlance, means that something is here-and-now. At the psychological level this is a description of the self—the sensation we all carry around with us that refers to the "I." Our sense of I-ness tells us that we exist; that we matter; that we count for something; that we are set apart from all others in the world by being *here* and not somewhere else, and by existing *now* and not at some other time.

For anyone who wants to understand the nature of prayer and something of the higher reaches of human religious and spiritual life, there is a great contradiction here. The goal of the spiritual life is to allow the highest and most divine qualities to flower in one's self. To a great extent this means leaving the self behind. For the self is local and God is not, and a call to God means moving away from the contracted, here-and-now sensation of self we call the ego.

All the major religious traditions have spoken plainly about the nonlocal nature of God. For example, the medieval Christian mystic, Meister Eckhart, stated,

> The more God is in all things, the more He is outside them.
> The more He is within, the more without. [5]

Or as Plotinus said almost two millennia ago,

> See all things, not in process of becoming, but in Being, and see themselves in the other. Each being contains in itself the whole intelligible world. Their

> All is everywhere. Each is their All, and All is each. Man as he now is has ceased to be the All. But when he ceases to be an individual, he raises himself again and penetrates the whole world. [6]

And from the Taoist tradition and the *Book of Chuang Tzu*, dating to approximately the turn of the fourth and third centuries B.C., we see the same theme:

> Do not ask whether the Principle is in this or that; it is in all beings. It is on this account that we apply to it the epithets of supreme, universal, total. . . . It has ordained that all things should be limited, but is Itself unlimited, infinite. . . . It is in all things, but is not identical with beings, for it is neither differentiated nor limited. [7]

Through all these observations, coming from various religious traditions of both East and West, there is the unmistakable message that the local, here-and-now condensation we call the personal self is not fundamental. There is something more that is essential to ourselves than the flimsy ego, and that whatever this essence is, it cannot be localized to a specific point in space and time. One of the most beautiful expressions of this all-pervading "something" that accounts for the nonlocal essence of both man and Nature was penned by the poet Yung-chia Ta-shih:

> One Nature, perfect and pervading, circulates
> in all natures,
> One reality, all-comprehensive, contains within
> itself all realities.
> The one Moon reflects itself wherever there is
> a sheet of water,
> And all the moons in the waters are embraced
> within the one Moon.
> The Dharma-body (the Absolute) of all the Buddhas
> enters into my own being.

And my own being is found in union with theirs. . . .
The Inner Light is beyond praise and blame,
Yet it is even here, within us, retaining its
serenity and fullness.
It is only when you hunt for it that you lose it;
You cannot take hold of it, but equally you
cannot get rid of it,
And while you can do neither, it goes on its own way.
You remain silent and it speaks; you speak and
it is dumb;
The great gate of charity is wide open, with no
obstacles before it. [8]

From the tradition of Islam, the great Persian poet Kabir echoed the same sentiment:

Behold but One in all things; it is the
second that leads you astray. [9]

Sometimes it is said that it is only mystically minded Orientals who hold to the vision of God in all and all in God. However, this vision permeates the Christian tradition as well. As an example, Saint Catherine of Genoa described her sense of oneness with the Almighty in explicit, almost shocking terms:

My Me is God, nor do I recognize any other Me
except my God himself. [10]

Again from the Christian tradition, Meister Eckhart left no doubt about his concept of the nonlocal process that envelopes both God and man:

To gauge the soul we must gauge it with God,
for the Ground of God and the Ground of the
Soul are one and the same.

And again from Meister's vision,

> The knower and the known are one. Simple people imagine that they should see God, as if He stood there and they here. This is not so. God and I, we are one in knowledge. [11]

And from the *Theologia Germanica* comes the same refrain of the nonlocal way in which God's goodness expresses itself—as something that is omnipresent, not distant and far away:

> Goodness needeth not to enter the soul, for it is there already, only it is unperceived. [12]

All the above quotations are taken from one of the most remarkable modern works on the eternal spiritual impulse of humankind, Aldous Huxley's *The Perennial Philosophy*. [13] Huxley's main concern is to show that there is a common thread running through the mystical urgings of humankind that bridges all the major traditions—the *philosophia perennis,* as Leibniz originally called it. He does not explicity refer to the nonlocal nature of the human mind, although he comes transparently close to the idea time and again, as in the allusions to the nonlocality of man's being that are expressed in the quotations above.

Because Huxley believed humanity's spiritual essence was in some sense eternal and thus complete from the beginning—that it was nonlocal in time—our spirituality was therefore not something that could be gained or developed across time. As he put it,

> The nineteenth century's mania for history and prophetic Utopianism tended to blind the eyes of even its acutest thinkers to the timeless facts of eternity. Thus we find T.H. Green writing of mystical union as

> though it were an evolutionary process and not, as all the evidence seems to show, a state which man, as man, has always had it in his power to realize. . . . in actual fact it is only in regard to peripheral knowledge that there has been a genuine historical development. . . . But direct awareness of the "eternally complete consciousness," which is the ground of the material world, is a possibility occasionally actualized by some human beings at almost any stage of their own personal development, from childhood to old age, and at any period of the race's history. [14]

Thus for Huxley the highest reaches of human experience, the spiritual, are not developmental and are therefore devoid of any history. This is a view of nonlocality in time, and it comes through again and again in his work in various ways.

Huxley was convinced that our concern over linear, flowing, historical time is a sickness, and that it resulted in our sense of alienation from the Divine. The habit of viewing ourselves as historical beings separated or "lost" from God was the result of fixing ourselves in time and space as isolated individuals, as separate egos set apart from all other beings and all other times. The whole point of the perennial philosophy was to break through this illusion of separateness and dualism. Put another way, it was to see through the falsity of the local view of ourselves. When this was done, the soul was recovered—not by "coming to" God through time, not by achieving some future state, and not by being rescued by the benevolent act of a god—but by waking up to the fact that, now and always, the soul and the Godhead are one and the same.

But the fixation on flowing time and history affects all disciplines, not just religion and spirituality. As an example, Huxley cites the views of the ethnologist Paul Radin, who observed that,

> Orthodox ethnology has been nothing but an enthusiastic and quite uncritical attempt to apply the Darwin-

> ian theory of evolution to the facts of social experience. . . . no progress in ethnology will be achieved until scholars rid themselves once and for all of the curious notion that everything possesses a history; until they realize that certain ideas and certain concepts are as ultimate for man, as a social being, as specific physiological reactions are ultimate for him, as a biological being. [15]

The point is not to investigate what ethnologists think about space and time, but to observe that in perhaps all areas of human endeavor it matters greatly how we choose to regard the world—whether as a collection of purely local events or whether as nonlocal. This choice has far-reaching implications in the lives of all persons, among which are the effects of prayer on health, as we saw above; in our concepts of the self; and in our religious and spiritual views. What is at stake is our commonsense attitudes toward space and time—all those assumptions that bind and fix us to specific locations and that generate the impression of a separate self.

Our problems assessing the role of prayer in health come about in large measure because of the curious way we think about the concept of distance and locality in medicine. Healers must be on site, close at hand, such as by the bedside or at the operating table. They cannot be across the county or beyond the state line, as in the above study. Because of the requirement of proximity, distant healing is an absurdity. Ergo, prayer is manifestly futile on the grounds of distance alone, of spatial separation, aside from the fact that it is nonmaterial and "mental." In medicine, then, locality has become an ironclad criterion: Long-distance therapeutic effects are the stuff of science fiction, not of real medicine.

The bugbear of locality is everywhere and infiltrates everything we do in medicine today. Because we insist that all therapies be physically based, we are off to the technological races, as we have been for over a century. And with arguable results—for, echoed in the experience of everyone, patients *and* physicians, there is an increasing awareness that something has been left out.

What has been omitted is the realization of who we are. Our therapies in medicine force us to assume the role of a purely local creature. But this is a false identity. It is a participation in hypocrisy, for it denies that we are nonlocal beings in space and time.

How Should We Pray? The Spindrift Experiments

If human consciousness does extend beyond the body through prayer, and if we want to use prayer to make things happen in the world, what is the best way to pray? *Is* there a superior way? Could one put the various methods of prayer to an objective test to find out?

Many people say no. It is wrong in principle, they claim, to try to "bring God into the laboratory." Many who claim to be healers have flatly refused to be tested objectively, stating that the scrutiny of observers under laboratory conditions interferes with their results. Because of such reasons the various methods of using prayer in healing have largely remained unexamined. By and large, each person goes his or her own way, using what seems to work best or following guidelines sanctioned by specific religious traditions.

Even in rigidly controlled studies that show the effectiveness of prayer, such as cardiologist Randolph Byrd's study in the coronary care unit of San Francisco General Hospital, described earlier in this chapter, the precise method of praying was not controlled and is not specified. The various prayer groups were simply told *to* pray, not *how* to pray. They were composed of Protestants and Catholics, and presumably they did not all follow the same method. The frequency and duration of prayers, the nature of the images held in the mind, and the specific goals of the prayers were left to the preference of each individual involved.

Ingenious attempts to assess objectively the effectiveness of various ways of praying have been quietly pursued for more than a decade by a unique organization called Spindrift.* ("Spindrift" is a variation of an old Scottish word describing sea spray driven by wind and waves. It suggests an interface of air and ocean, of the ethereal and the concrete, of mind and body.) Spindrift researchers use very simple tests and minimal equipment, yet their experimental designs are intrepid and their results are open to the critical inquiry of anyone. Taken as a whole, their experiments are extremely helpful in filling in the enormous gaps in our knowledge about how prayer works.

A central assumption made by the Spindrift researchers is that all humans have "divine attributes, a qualitative oneness with God." [16] This assumption is a central thesis of this book: There is a nonlocal quality of human consciousness; consciousness, like the Divine, is infinite in space and time and is ultimately One.

The first questions asked by the Spindrift researchers are fundamental. Is spiritual healing real, does prayer work, is there an effect that can be measured, and is the effect reproducible? One of the simplest ways of answering these questions is to test the interaction of a healer or prayer-practitioner with a simple biological system such as sprouting seeds. If the healer prays for one batch of germinating seeds and not another, is there a difference in the rate of germination?

In one test, rye seeds were divided into two groups of equal number. They were placed in a shallow container filled with vermiculite, a light, soil-like substance commonly used by gardeners. A string was placed down the middle of the container, dividing the seeds into sides A and B. The seeds on one side were prayed for and the others were not. After the seeds had grown, the slender rye shoots were counted. Results consistently indicated that there were significantly more rye shoots in the

---

*Details of the experiments described here as well as further information about Spindrift can be obtained by writing the organization at 2407 La Jolla Dr. NW, Salem, OR 97304.

''treated'' (prayed-for) side than in the control side. This simple test, repeated many times with many practitioners, indicated that the effect of thought on living organisms outside the human body was significant, quantifiable, and reproducible; and that the effects of human consciousness are not confined to the brain and body.

But in ''real life'' we pray for unhealthy persons as well as for healthy ones. So, the researchers asked, what if the rye seeds being prayed for were unhealthy instead of healthy? Would the prayer continue to work? To test this question Spindrift researchers stressed the rye seeds by adding salt water to the seed container, keeping the rest of the above experiment the same. The salt water diffused upward through the vermiculite, eventually reaching the seeds. The results of prayer were now even *more* striking: The ratio of the treated (prayed-for) to control (not-prayed-for) shoots increased sharply, indicating that prayer worked *better* when the organism was under stress.

What if the stress on the seeds was increased? Would prayer still work? The Spindrift researchers ran the same experiment several times, each time adding an extra measure of salt to the water in the bottom of the seed container. With each added increment of salt the effect of prayer was increased. The saltier the solution bathing the seeds, the more seeds germinated when prayed for. This suggested that prayer works best when physical conditions are worse instead of better. (There is a strong parallel in clinical medicine to this finding. We know, for instance, that placebo pain medication—a ''sugar pill,'' which has no known biological effect—works better with severe pain than with mild pain.)

Then the researchers changed systems. When soy beans were used instead of rye seeds, and when temperature and humidity were used as the stressors instead of salt water, the same results were seen: Prayer worked best with increasing stress on the organism.

Then the experimenters asked another logical question: Does it matter how *much* one prays? If an individual prays ten minutes versus twenty minutes, will the effect be the same or different? A simple test of this question involved four containers, this time using soy beans. One container was marked "control" and was not prayed for. The other three were marked X, Y, and Z. In each run of the experiment the X and Y containers were prayed for as a unit, and the Y and Z containers were also prayed for as a unit. This meant that the Y container was prayed for twice as much as the X and Z containers. Germination measurements indicated that the Y container did *twice* as well as the X and Z containers, showing that the measurable effect was in proportion to the amount of prayer given, twice as much prayer yielding twice the effect.

Since the Spindrift experiments involve a control group and a prayed-for group, a logical question seemed to be, How does the prayer "know" which seeds to help? In search of an answer, experiments were conducted in which the person praying was kept uninformed about the nature of the seeds he was praying for. Results showed a drastic reduction in the effect of prayer. Thus, the researchers concluded, the more clearly the practitioner is aware of his subject, the greater the effect of his prayer. "In order for our prayers to have any effect," they concluded, "we need to know who or what we are praying for." [17]

Are some healers better than others? It proved quite easy for the Spindrift researchers to test the relative effectiveness of healers. In the type of experiments described above, as well as in sensitive tests with yeast that measure the amount of carbon dioxide produced by the yeast culture, it proved to be the more experienced practitioner rather than the inexperienced one who produced more powerful outcomes.

One of the more remarkable observations of the Spindrift researchers is that there is no loss of effect as the number of parts involved increases. In tests with seeds, for instance, the results were comparable no matter whether the total number of seeds was large or small. Thus, after many years of research, the Spindrift researchers have formulated the *law of the conceptual whole:* So long as the practitioner can hold in his mind an overall concept of the system involved, the effect of prayer is constant over all components.

One of the most important contributions made by the Spindrift researchers is the distinction between "directed" and "nondirected" prayer. Directed prayer occurs when the practitioner has a specific goal, image, or outcome in mind. He is directing the system, attempting to steer it in a precise direction. In healing, he may be praying for the cancer to be cured or the pain to go away. In the seed germination experiments above, he is praying for a more rapid germination rate. Nondirected prayer, in contrast, uses none of these strategies. It involves an open-ended approach in which no specific outcome is held in the imagination. In nondirected prayer the practitioner does not attempt to tell the universe what to do.

Which technique—directed or nondirected prayer—is more effective? Is prayer more powerful if a specific goal is held in the mind, or does a simple, "Thy will be done" approach work better?

The Spindrift tests are unequivocal. Although both methods were shown to work, the *non*directed technique appeared quantitatively much more effective, frequently yielding results that were twice as great, or more, when compared to the directed approach. This may surprise persons who favor the techniques of directed imagery and visualization that are quite popular today. Various imagery schools contend that if the cancer, for example, is to be cured, a specific image must be employed as to how the end result will come about. Some studies have suggested that the more robust and aggressive the image is, the better the outcome. But Spindrift's quantitative tests say otherwise.

In one particular experiment, directed and nondirected prayer were put to the test. The organic system involved was a mold growing on the surface of a rice agar plate, the kind that bacteriologists and mycologists routinely use. The mold was stressed—washed in an alcohol rinse so as to damage it and retard its growth, but not enough to kill it. A string was then placed across the mold, marking it into sides A (the control side) and B (the treated or prayed-for side). When directed prayer was used to encourage the growth of side B, nothing happened; growth remained static. But when directed prayer was replaced by nondirected prayer, in which no goal was outlined in the mind of the healer, side B began to multiply and formed additional concentric growth rings.

As a result of numerous tests on a variety of biological systems, the Spindrift researchers suggest that the healer strive to be completely free of visualizations, associations, or specific goals. Physical, emotional, and personality characteristics should be excluded from thought and replaced by a "pure and holy qualitative consciousness of whoever or whatever the patient may be." [18] It is this method that they refer to as genuine spiritual healing. Methods that rely on directed prayer, in contrast, are referred to as "psychic" healing, "faith" healing, "mental" healing, or the placebo effect.

In a remarkable series of experiments Spindrift researchers have demonstrated the ability of prayer to influence random processes in nature. In a germination experiment they put 500 stressed mung beans, oversoaked in salt water, in each of three paper cups with holes punched in the bottom. The cups were marked either H for "heads," T for "tails," or C for "control." Then a penny was put in a box and shaken thoroughly, and the box was not opened. Treatment in the form of prayer was then given daily for a week to the H and T cups, and none to the C cup. All the while the practitioner directed the effects of his prayer to correspond to the upturned side of the penny, without knowing whether it was actually heads or tails. At the end of a week there were twice as many sprouted mung beans in the "heads" cup as the other two; and when the box was opened the penny also was "heads." A mere coincidence? To find out, a series of tests of this sort was done, with consistent findings. Using six or seven cups the Spindrift researchers found they could call correctly the face of a die, predict the number of a playing card, or the denomination of a bill in an envelope, depending on how the test was set up. [19]

One of the problems with directed prayer is that the practitioner or patient frequently does not know which image is best. Should he or she pray for an increase or decrease of blood flow to a specific organ? For an increase or decrease of a particular type of blood cell? These questions can be bewildering to persons who want to steer their physiology and their health in a particular direction. The Spindrift experiments are consoling on this point. They suggest that it isn't necessary to know which way the body ought to go. One need only to pray, they say, for "the norm" to happen, asking only for "what's best"—the "Thy will be done" approach. This is particularly obvious in a series of germination experiments in which the practitioner did not know what was best for the seeds involved. One batch of the seeds was oversoaked and thus heavier than they should have been for proper germination to occur; another batch was undersoaked and lighter than optimal. The seeds were being evaluated early in the germination process according to changes in weight (properly germinating seeds gain weight early in germination). Ideally, the oversoaked seeds should have eliminated excess water early and become lighter, and the undersoaked seeds should have absorbed water and become heavier. But not knowing which batch was which, which change was the practitioner to pray for? He was in the same position as the patient who is uncertain about which image to employ to change his physiology. So, rather than "telling the beans what to do," the practitioner used nondirected prayer and trusted that the beans would simply move toward the norm according to what was best for each seed. The nondirected approach worked. The results showed that the oversoaked beans eliminated water and lost weight in the early germination phase, and the undersoaked beans gained water and increased their weight. These experiments suggest that nondirected prayer moves organisms toward those states of form and function that are best for them, and that the practitioner need not know what "best" is.

The Spindrift experimenters are aware of the scientific heresy implied by their findings. "Scientifically," they state, "it is a shocking thing to think of 'force' as intelligent, loving, kind, good, and aware of needs. But in each test prayer, somehow linked to a loving intelligence, moved the seeds toward their norms. When prayer was for seeds in different conditions—the same prayer at the same time—[the outcome] was always in the direction that was best for the individual need of the individual beans. . . ." [20]

It is difficult, however, to put nondirected prayer into practice. When health fails, when problems arise, it is our own preferred outcomes that we usually pray for. We believe we *do* know what the norm should be, and we waste no time in telling the universe how to behave. The tumor should vanish, the pain should subside, we should become more prosperous: All this we claim to know in advance.

But a moment's reflection should tell us that nature could not endlessly tolerate this strategy. If all human prayers for restored health were answered, almost no one would die, and our planet would have become overpopulated long ago and would be unfit today for human life. It seems written into nature: We should *not* pray for a divine bailout with every catastrophe. Common sense tells us there is a time when death is natural, when it is the norm.

But when? When is it right to pray for healing, and when is it improper? Given our limitations, perhaps we can never know with certainty in any given situation what is best to pray for. Is it then not better to follow a *non*directed form of prayer, relying on a higher Universal Intelligence to make the decision and to decide what is the norm for our particular exigency? *Especially* in view of the fact that nondirected prayer appears more potent anyway?

One of the great figures in the history of science who had an abiding faith in "the norm"—the intrinsic harmony and rightness of the universe—was Albert Einstein. This awareness evoked in him what he called a "cosmic religious feeling." As he put it, "The individual feels the futility of human desires and aims and the sublimity and marvelous order which reveal themselves both in nature and in the world of thought. . . .[A] person who is . . . enlightened . . . has . . . liberated himself from the fetters of selfish desires and is preoccupied with thoughts, feelings, and aspirations to which he clings because of their *super*personal value." (Emphasis added) [21] It is just this awareness of our own limitations and fallibilities, and the trust in the harmonious workings of the universe, that provide the impetus for the approach of nondirected prayer.

As an intern working in a busy emergency room, I encountered a piece of advice that embodies this trust in "the norm" and which I shall never forget. Frequently our resuscitation attempts on dying persons would prove unsuccessful. After chest massage, defibrillation, and a variety of drugs had been administered to the patient, all to no avail, I would turn helplessly to a particular senior resident physician and implore him about what to do next. To my consternation he would usually reply, "Just do the right thing." This bit of gallows humor contained more wisdom than I knew at the time. Sometimes we simply cannot discover what the right thing is; and we must acknowledge that eventually the right thing, the norm, will *not* mean the prolongation of life or even a temporary improvement in illness, but death.

This seems harsh and painful as long as we identify ourselves solely as isolated individuals possessing single minds that are locked in linear time, moving toward extinction. But when we go beyond the local definition of ourselves to an awareness of our nonlocal Self, the situation changes. Aware of our infinitude in space and time, we are then free to contemplate the divine qualities that each of us contains. Within this awareness we are poised to understand the wisdom of the norm, even though it may mean the extinction of the physical body or the progression of incurable illness.

Becoming aware of our nonlocal nature leads us to an understanding of the wisdom of nondirected prayer—that it makes more sense to pray "*Thy* will be done" than "*my* will be done." To pray, in other words, for the right thing.

Nonlocal Healing

The healer does not "do" or "give" something to the healee; instead he helps him come home to the All, to the One, to the way of "unity" with the Universe, and in this "meeting" the healee becomes more complete and this in itself is healing. [In] Arthur Koestler's words: "There is no sharp dividing line between self-repair and self-realization."

—LAWRENCE LESHAN [22]

A person is neither a thing nor a process, but an opening or clearing through which the Absolute can manifest.

—MARTIN HEIDEGGER

As a physician, I have had many experiences over the years that have led me to conclude that the world of clinical medicine is truly bizarre and unpredictable, a territory where almost anything is possible. Most of my colleagues, I feel, agree, for almost all physicians possess a lavish laundry list of strange happenings unexplainable by normal science. A tally of these events would demonstrate, I am convinced, that medical science has not only not had the last word, it has hardly had the first on how the world works—especially when the mind is involved.

Sometimes, when the setting is right, the private beliefs of physicians about "mind cures" come to the surface. Several years ago I was asked to address the Department of Internal Medicine at my hospital on biofeedback at the regular monthly meeting. (For years I had directed a biofeedback clinic in addition to my practice of internal medicine.) I agreed with great reluctance because I sensed a strong hesitation on the part of the local body of physicians to refer patients for biofeedback therapy, which I took to be evidence of a closed-minded attitude toward this practice. I frankly was not eager to stir up a "mind or matter" debate in a formal meeting, and did not want to be subjected to the hostility and disapproval of my peers. Moreover, I suspected that few physicians would even attend the presentation. But my concerns proved groundless. The early evening meeting was packed, standing room only, the biggest crowd in years. My colleagues were not only cordial, but seemed fascinated by the topic. There was a long period of discussion following my talk, during which intelligent, incisive questions flowed freely. I was amazed, and was somewhat ashamed of my timidity and hesitation over exposing myself to their criticism, which never materialized.

But the biggest surprise came after the lecture. As I was leaving the area several physicians, all respected internists, followed me into the hallway for conversation. Each of them began in essentially the same way: "If you think the cases *you* just presented were weird, wait until you hear about *this!*" whereupon the doctor would reveal to me some bizarre incident involving a patient of his, a story that he was obviously hungry to share. Apparently my lecture on biofeedback had displayed my openness to the unusual and the uncommon, and my colleagues suddenly sensed they had an ally to whom they could divulge these stories that they had been hoarding for years. But they were still squeamish, and as they spoke one or two of them glanced furtively about as if concerned that the wrong colleagues might be listening. All their accounts had to do with healings that were completely unexplained or incurable illnesses that simply "went away." The implications of all these cases were the same: It was the mind that had somehow made the difference.

That event left an unforgettable mark in my memory. My

colleagues had given me an insight into their private experiences, if only for a few minutes. Over the years, many similar instances have followed, revealing an increasing openness about the role of the mind in healing that I did not anticipate. As a result, I am convinced that the time is right to take a fresh look at psychic healing, no matter how heretical it has been in the past.

Just as prayer can be effective at a distance, so can psychic healing, although it is frequently used in close proximity to the patient (which does not make it any less extraordinary). Many of the great psychic healers attest to a striking feeling of nonlocality surrounding the healing event. A compelling sense of oneness and unity with the world and all in it, including the patient, seems to take over for many of them. The descriptions they give of their healing efforts leave little doubt that they do not experience themselves as fixed to a specific point in space and time. Going beyond the sense of self, they fuse with the world outside them, leaving their personal sense of identity behind.

One of the most incisive analyses of the processes involved in psychic healing is the brilliant work of psychologist Lawrence LeShan. As psychophysiologist Jeanne Achterberg, author of *Imagery in Healing,* [23] points out, when one begins to investigate an area of medicine in which the interaction of mind and body plays an important role, one frequently finds that LeShan has already been there years before, doing the groundwork before anyone else arrived. LeShan's book, *The Medium, the Mystic, and the Physicist,* [24] is of inestimable importance in placing psychic healing in a rational context that is approachable by science. The following discussion owes much to LeShan's insights.

The most important step in understanding psychic healing is to appreciate the underlying nonlocal nature of reality in which healers operate. This is, of course, not the way we see the world in everyday life. The ordinary way of perceiving things is entirely local: We "know" that all events, especially our bodies and minds, occupy separate places in space and points in time. But throughout history human beings have broken through to a nonlocal reality in which all of space begins to appear as a seamless garment, totally unitary, and in which all of time appears to be undivided into past, present, and future. W. T. Stace, a great scholar of mysticism, described this way of experiencing the world:

> The whole multiplicity of things which comprise the universe are identical with one another and therefore constitute only one thing, a pure unity. [25]

Or, as echoed through the Hindu tradition in the *Mandyyuka-Rarita Upanishad,*

> All objects are in origin unlimited like space
> And multiplicity has no place in them in any sense. [26]

The well-known psychiatrist, Kurt Goldstein, described both these ways of seeing the world—the everyday and the mystical—as essential to man's nature. Importantly, the nonlocal way of knowing the world is not simply an artifact of experience known only to overtly mystical types. As Goldstein says,

> I have come to the conclusion that man always lives in two spheres of experience: the sphere in which subject and object are experienced as separate and only secondarily related, and another one in which he experiences oneness with the world. . . . Because we observe these experiences in normal human beings, we must say that they, and the world in which they appear, belong to man's nature. [27]

The unitary way of seeing is not aberrant. Too often those who describe this state have been accused of having some sort of mental derangement such as schizophrenia by "experts" who have no personal knowledge of this way of being in the world, and who thus cannot be expected to understand it. "Depersonalization," "loss of identity," or "breakdown of boundary" are some of the pejorative descriptions of this mode of perception. Prescriptions of antipsychotic medication and even hospitalization of the "affected" individual are frequent. But this kind of experience, of itself, is not pathological if Goldstein is correct, and if we are not to condemn as mentally unbalanced the best part of our most gifted poets, writers, sculptors, musicians—and our psychic healers.

LeShan terms this mode of seeing the world the Clairvoyant Reality, because many seers, or clairvoyants, describe this as the state in which they function when they gain information in paranormal ways: seeing the future, knowing distant events, and others. However, I want to call this state *Nonlocal Reality* in order to emphasize its difference from the local, ordinary way of seeing the world.

We have seen that Nonlocal Reality's most prominent feature is a central unity to all things and events. In addition, time is seen in a nonordinary way. Past, present, and future are illusions we project, and are not fundamental. Within this context of connectedness, evil is mere appearance because it, like everything else, is part of the whole and is related to the good. When we can genuinely enter this understanding (a term that originally meant "to stand under" or "to be a part of"), these irreconcilable opposites reveal their unity to us. Moreover, we begin to see that there is an alternative way of gaining this information than through the senses. [28]

How does psychic healing work? As LeShan's research shows, the largest group of healers regard their work as "prayer" and believe their effectiveness is due to the intervention of God. The next largest group ascribe their work to "spirits." For them the healing comes about after they have set up some special channel between the spirits and the patient. And a third group believes they are transmitters or originators of some form of "energy" that has healing effects.

How can we get a handle on these claims? How are we to deal with statements about God, spirits, and energy? As far as the first group is concerned, I agree with LeShan in not wanting to oppose the hypothesis that God is doing the work. Yet we want to go further. We need to know just *how* God is working so we might systematically and more reliably make use of the effect. As for the second group's claim that spirits underly their work, we are at a loss to either prove or disprove their contention. There has been no convincing scientific evidence for the existence of spirits, in spite of the valiant and unflagging efforts by many researchers. We can allow that spirits may indeed be what they claim to be; but they may also be a split-off aspect of the personality of the healer, as in the multiple personality syndrome. Some reputable healers, in fact, have confessed their inability to rule out this possibility. Almost invariably spirits are given names, and sometimes they are regarded as having well-developed personalities, and occupying places or points in time and space. Many healers' descriptions may be metaphorical at best, some hopelessly anthropomorphic, like the pictures the early anatomists used to make of man's innards—levers, gears, belts, and pulleys stretching from head to toe.

People in the fringe areas of medicine are not the only ones who have toyed with the existence of spirits. Sir William Osler, the father of modern medicine, revealed some of his thoughts about them in a bit of verse contained in "Science and Immortality," his Ingersoll Lecture given at Harvard in 1904. Sir William acknowledged, after commenting on "the futile search of science" for discarnate entities, that,

> Perhaps they live in the real world, and we in the shadow-land! Who knows? Perhaps the poet is right:
>
> I tell you we are fooled by the eye, the ear:
> These organs muffle us from that real world
> That lies about us; we are duped by brightness.
> The ear, the eye doth make us deaf and blind;

> Else should we be aware of all our dead
> Who pass above us, through us, and beneath us.

If the great Osler had serious thoughts about the spirit world, perhaps his words can encourage those physicians and others skeptical of psychic healing to reconsider.

What about the "energy" group of healers? The problem is that no measurable form of energy has ever been discovered in psychic healing (although this does not mean it does not exist). Is it too subtle to be detected by current scientific devices? Perhaps; no doubt we have not discovered all the forms of energy that exist in the universe. But if there is any energy involved, it would seem completely unlike any form of energy we know today. For one thing, the "energy" claimed by psychic healers has been shown to be capable of penetrating all known barriers such as brick walls and traveling great distances without being diminished, as in the prayer study in San Francisco General Hospital described above. The energy explanation seems to violate the tenets of nonlocal reality. As LeShan points out, energy flows or travels between two "things." But the thinglike nature of the world is deemphasized in the nonlocal mode of being, which is the experiential mode of most psychic healers.

For most healers, entering the state of Nonlocal Reality in which they actually do their work involves participating in an altered state of consciousness that is very similar to meditation or prayer. In this state they see themselves and the patient as one, as the subject-object bridge is completely overcome. As Edgar Jackson, a famous psychic healer put it, "Prayer as a specialized state of consciousness moves beyond the usual consideration of real or unreal, conscious or unconscious, organic or inorganic, subjective or objective to a place where [the healer] is dealing with the totality of being at one and the same time in a way that produces sensitivity to the whole." [29]

This is a striking description of an embracing unity between the healer and the patient. Separateness takes a back seat and oneness comes to the fore in this mode of experience. Individuality and ego are simply not a part of the healer's repertoire, in sharp contrast to many of the TV evangelist-healers who virtually drip with "personality," "power," and "presence."

The Importance of Love in Healing

But the experience of Nonlocal Reality itself does not completely explain this state. Something else is needed to give life to the feeling state of nonlocality. After all, studies have shown that infants seem to be "at one" with their environment, and so do many psychotics and drug takers. What is to separate their nonlocal experience from that of healers? A major difference is that the unity that the healer feels with the patient is infused and transformed by love and caring. This is an essential factor whose importance simply cannot be overestimated.

> In Agnes Stanford's words, "Only love can generate the healing fire." Ambrose and Olga Worrall have said, "We must care. We must care for others deeply and urgently, wholly and immediately; our minds, our spirits must reach out to them." Stewart Grayson, a serious healer from the First Church of Religious Science, said, "If this understanding is just mental it is empty and sterile" and "the feeling is the fuel behind the healing." Sanford wrote: "When we pray in accordance with the law of love, we pray in accordance with the will of God." [30]

Love and caring: these have forever separated this form of healing from the exercise of mere technique. Nothing is gained from simply entering a state of nonlocal consciousness. One can do as much with drugs, alcohol, sexual intimacy, or holding one's breath. Something more is needed, that which is contained in the adage, "The secret to patient care is caring for the patient." Modern medicine, because it has tried to function completely objectively, has become care-less. As a consequence, its power to heal has atrophied. Without the catalyst of love and caring, medicine becomes a mere manipulation of tissue, an orchestration of chemistry.

This is one of the most important lessons orthodox medicine could learn from the psychic healers. It is indeed strange today that love on the part of the healer, if it is an essential ingredient, is never mentioned in our medical schools, although this may be changing, given the enthusiastic reception of *Love, Medicine, and Miracles,* written by Dr. Bernard Siegel, the Yale surgeon. [31] Yet, the most elaborate treatment algorithms—the flow diagrams in medical textbooks that show the next step in the treatment of an illness—never contain a box that says "LOVE!" or "CARE!" And that is why the algorithms fail, although they appear to be foolproof and to take every eventuality into consideration. And that is also the reason why computers, flow diagrams, and all decision-making charts can never duplicate the job of the healer.

When psychic healing takes place, the experience of love and nonlocality are so tightly connected that they cannot be teased apart. Healers experience a sense of *holiness* with that of *wholeness,* such that the universe seems Godlike, enchanted, and filled with love. As the Worralls expressed this awareness, "In true prayer our thinking is an awareness that we are part of the Divine Universe." The psychic healer does not have to deliberately strive to add the element of love to the healing agenda; rather, the love is endemic to the genuine nonlocal sense of wholeness. We are not speaking here about love as an emotion. Rather, this greater love is more akin to a deep, inner desire to reestablish completeness in the patient, the loss of which has expressed itself in the form of the illness. Love for the patient *is* this yearning to help him or her be whole, to be One. Without this love healing simply does not occur.

The scriptures tell us that God is love. God, too, is the One, the Pure Unity outside of which nothing can exist. And the psychic healers frequently are God-conscious people who are deeply sensitive to this connecting triad: God, Love, Unity. This is why much of psychic healing is consistent with the great spiritual visions of the world's great religious creeds; and why, too, so many of the greatest healers have performed their services within the context of a formal religion. As the great medieval Christian mystic Jan van Ruysbroeck said,

> The image of God is found essentially and personally in all mankind. Each possesses it whole, entire and undivided, and all together not more than one alone. In this way we are all one, intimately united in our eternal image, which is the image of God and the source in us of all our life. [32]

Another Type of Healing

LeShan describes another type of healing that is experientially quite different from the "unity" healing above, and which occurs less frequently. In this less common form the healer has the feeling of remaining separated from the patient. The healer feels a flowing energy emanating from his or her palms. These are placed on either side of the pathological area of the patient, and the "healing energy" is allowed to "pass through" the site. In perhaps half of these instances the patients feel as if heat were actually passing into their bodies in spite of being totally unfamiliar with what to expect from the experience. Others describe some type of "activity" in the affected part, while still others, a distinct minority, describe a cold sensation.

In this type of healing, the healer actively *tries* to heal by manipulating the "energy flow." This may sound as if the healer is back in the old familiar local world that is populated once again with separate things isolated in space: the healer, her hands, the patient, the patient's affected area, and the energy that is being manipulated. But a forgetting of the self takes over to a great extent. As healer Harry Edwards explained:

> Only one thing exists and that is [the] hands through which the healing power flows. The healer's whole being is concentrated on his fingers—nothing else seems to exist. The desired result is the only thing that is his concern. [33]

Where does the energy come from? Some healers believe the origin of the healing energy is themselves. Others believe the source is God, in which case they are simply trying to transmit it to the patient—all of which is different from the Nonlocal Reality described earlier. In fact, it is *so* different that LeShan considers it a possible cop-out in which the healer and the patient say, in effect,

> We are both too frightened of all this closeness and uniting that is a part of psychic healing. Let's pretend with each other that all that is happening is a flow of energy [that] is coming out of the hands and treating the problem. That way we will both be more comfortable. [34]

But on one point the practitioners of these different types of healing seem to agree totally: love and caring are essential if the healing endeavor is to succeed.

In summary:

• In spite of the fact that skeptics are rarely persuaded that psychic healing exists (improvements in illness can always be attributed to "the natural course of the disease"), dramatic healing events do occur that do not involve interventions with drugs or surgery but which are mediated through the efforts of healers.

• These extraordinary events do not occur in any sort of locally perceived reality, but in a *nonlocal* mode of being as experienced by the healer.

• Modern medicine may be limiting its effectiveness by its chronic insistence that the local, here-and-now reality is the only reality there is, and that this is the only reality in which healing can occur.

- Psychic healing, by the very fact of its existence, gives us strong evidence that there is something of our psyche that is soul-like, nonlocal in space and time. It shows us that our time is more than the moment and even more than a single lifetime, and that our place is not just where we are now, but everywhere.

Miracles Do *Happen: A Case History*

On June 24, 1981, an apparition of the Virgin Mary appeared in the little village of Medjugorje in Yugoslavia. Since then it has been seen every day, and as a result Medjugorje has become a pilgrimage site for millions all over the world.

It has long been observed that extraordinary healings take place in association with apparitions such as the one seen in this small Yugoslavian village. In fact, several hundred remarkable cures have been reported from Medjugorje. Father Slavko, a Franciscan monk with a Ph.D. in psychology, who lives and works there, states that he can sometimes tell who will be healed. "It's very often the people who come and don't determinedly *want* healing who are affected," he says. "They come with an open mind and ask for healing but they have not come with this as the single-minded purpose of their trip." [35]

Brendan O'Regan, vice president for research at The Institute of Noetic Sciences in Sausalito, California, visited Medjugorje to investigate these strange happenings, and agrees there is an interesting psychological profile of those being healed. "These people *are* in a very different place psychologically, emotionally, and indeed psychophysiologically," he observed. "There is a sad, faraway look in [their] eyes . . . that is unmistakable. It seems like a kind of yearning for something, the search for a memory, the need for an all-embracing experience of love . . . ." [36]

O'Regan and his Institute are in the process of analyzing the data available worldwide on spontaneous cures such as those reported from Lourdes, Medjugorje, and elsewhere. This information has never been systematically examined before. So far the project has combed over 860 medical journals and over 3,000 individual articles—the largest compilation in the world. Their report will be entitled *Remission with No Allopathic Intervention,* and should go far to establish the legitimacy and commonness of so-called miracle cures.

If we look at the actual case histories of sick persons who undergo unexpected cures, the observations of Father Slavko and O'Regan do seem to hold true. In them we can get the feeling that a "nonlocal mode of being" and the "nondirected" approach, in which one simply asks that the "right thing" be done, do seem to play a crucial role in getting well. At first glance it might seem impossible to put these concepts into practice. Serious illness, after all, focuses the attention on the body and the here-and-now: the local mode of being. How could one conceivably step outside it? And how could one genuinely adopt a nondirected approach to illness? How is it possible to be neutral about the outcome of a major disease process that may result in death? How could one avoid steering the outcome of any treatment in a specific direction?

Perhaps we can never fully understand how this might be possible without personally experiencing a "miracle event" ourselves. But we can look at the experiences of others and try to put ourselves in their place—such as the case of a middle-aged Italian, Vittorio Micheli. According to the official report of the Medical Commission of Lourdes, France, which investigated his case, he was admitted in 1962 to the Military Hospital of Verona, Italy with a large mass in the left buttock region. This mass was so extensive it limited the normal range of movement of the left hip and was excruciatingly painful. X rays showed extensive destruction of the bones of the pelvis and hip, revealing that the mass was literally eating away Micheli's skeleton. At the end of May a biopsy was taken from the mass, which by this time was quite extensive. Pathological examination showed the lesion to be a fusiform cell carcinoma, an aggressive, destructive form of cancer. Micheli was by now very sick. At this stage his leg was immobilized in a frame from his pelvis to his feet, and he was sent during June for radiation therapy at a distant medical center. For some reason he was discharged four days later without receiving any irradiation treatments at all, and went to the Military Hospital at Trente, where he was admitted.

During the next ten months Micheli received no specific treatment at Trente. X rays showed the progressive destruction of bone in the pelvis and hip, and he developed increasing loss of all active movement of the left lower extremity. By now the entire hip had been eaten away, and the man was literally falling apart. His leg was detaching from his body as the supportive structures of the hip and pelvis—cartilage, tendons, and ligaments, in addition to the bone—were destroyed by the cancer. Then on May 24, 1963, approximately a year after his original diagnosis, he left for Lourdes where he was bathed several times, still in his plaster cast.

The report of the Medical Commission of Lourdes states that Vittorio Micheli was in a serious condition when he came to the shrine. He had lost vast amounts of weight and could not eat. But on being bathed his appetite returned immediately, he felt a resurgence of energy, and he had strange sensations of heat moving through his body. His friends then took him back to the hospital, where he began to gain weight and increase his activity levels. About a month later the doctors removed the cast for more X rays, and discovered that the cancer had decreased in size. Then, under their observation, the tumor disappeared completely within a few weeks. But another event happened that was even more amazing than the disappearance of the tumor: The bone of the pelvis, hip, and femur began to regrow, and with time *completely reconstructed itself!*

Two months after being bathed at Lourdes, Vittorio Micheli went for a walk.

The physicians of the investigative commission expressed the significance of these events quite clearly. They stated,

> A remarkable reconstruction of the iliac bone and cavity [of the pelvis] had taken place. The X rays made in 1964–5, –8 and –9 confirm categorically and without doubt that an unforeseen and even overwhelming bone reconstruction has taken place of a type unknown in the annals of world medicine. We ourselves, during a university and hospital career of over 45 years spent largely in the study of tumors and neoplasms of all kinds of bone structures and having ourselves treated hundreds of such cases, have never encountered a single spontaneous bone reconstruction of such a nature. [37]

Reading between the lines of the report, one can detect the astonishment of these physicians. They further stated,

> Definitely a medical explanation of the cure of sarcoma from which Micheli suffered was sought and none could be found. He did not undergo specific treatment, did not suffer from any susceptible intercurrent infection that might have had any influence on the evolution of the cancer [sometimes infections appear to stimulate the body to reject a cancer].
>
> A completely destroyed articulation was completely reconstructed without any surgical intervention. The lower limb which was useless became sound, the prognosis is indisputable, the patient is alive and in a flourishing state of health nine years after his return from Lourdes. [38]

There is no test that can tell us the mental state surrounding Vittorio Micheli's cure. Was he absorbed in a local or nonlocal perception of himself? Did he expect to be cured? We do not know. That is why the Spindrift studies we examined previously are so important. In them, one *can* show which forms of mental intent work best on shaping the health of simple biological systems.

Neither do we really know *why* he recovered. Whether any single ''miracle'' case is the result of psychic healing, prayer, a particular mental state employed by the patient, or whether, as skeptics always maintain, it is a case of ''spontaneous regression,'' may be impossible really to know. The crucial thing is to admit that, regardless of the seriousness of any illness, unexplainable and radical cures are possible. But how *likely* are they? If such cases are exceedingly rare, perhaps we should not emphasize them for fear of giving false hope to sick persons. Yet it is just as unethical to give ''false negative hope,'' as Dr. Rachel Remen has put it, wherein we paint an overly gloomy outlook to patients. Despair and a sense of doom can have devastating and fatal effects on health. [39] One way to avoid giving ''false negative hope'' is to admit freely the possibility, no matter how infrequent, for any disease to disappear totally and suddenly, sometimes in violation of all known physical principles.

''The mind,'' Dr. Jonas Salk has said, ''in addition to medicine, has powers to turn the immune system around.'' [40] In closing this section on prayer and psychic healing, I want to emphasize Dr. Salk's words ''in addition,'' which suggest that complementary forms of therapy are required. *We are compound creatures*. We have not only minds that can affect our bodies, but also bodies that *do* respond to the physical effects of drugs, surgery, irradiation, and similar interventions. The benefits of physical approaches to treating human illness are real and undeniable, and it would be foolish to abandon them in favor of an approach based completely on the mind.

Today there is tremendous ambivalence and confusion on this issue. With the discovery that the mind—emotions, attitudes, feelings, and perceived meanings—has a major role to play in illness, many persons leap to the conclusion that, in health and illness, ''it's *all* mind.'' This attitude is a perversion of medical reality. Not only can it lead to inhumane approaches in treating illness, it can cause a sense of guilt and shame when one falls ill. [41] What we need is a balanced approach, one that takes *all* of human nature into consideration when illness arises, the mental *and* the physical. The body *can* be regarded as a local phenomenon, and it can be treated with local, physically based approaches anchored in the here-and-now. But there is another aspect of our being that is *nonlocal*—the mind—that is infinite in space and time, and which, although nonmaterial, can also bring about profound changes in the state of the physical body. Only by combining both approaches can we discover the best therapeutic reality.

3

Bodymind

There is a measure of conscious thought throughout the body.

—THE HIPPOCRATIC WRITINGS

Whisperings of the Blood

I have been and still am a seeker, but I have ceased to question stars and books; I have begun to listen to the teachings my blood whispers to me.

—HERMANN HESSE
Prologue to *Demian*

The blood speaks. Hesse was correct, as are all the poets, writers, and seers who, throughout the ages, learned to listen to the whisperings of their blood. The mind is in the blood, just as Hesse implied. This is not a poetic statement; it is not a metaphor; it is a literal truth.

Today we know that certain cells in the brain make various chemicals that affect our mood and how we think. Depression, agitation, tranquility, fear, anxiety—these and many more common emotions are decisively linked to the actions of certain chemicals called neuropeptides and neurotransmitters. Dozens are now known to exist, and it is certain that many more will be discovered. Many cells in the blood make chemicals that in some instances are identical to those made by brain cells. Among these are the famous endorphins, chemicals that are potent, naturally occurring pain relievers that also affect our moods. Blood cells also contain receptor sites for various hormones and chemicals identical to those in the brain.

The picture that begins to emerge is that there seems to be more than one brain. There is the *anatomic* brain, which is the brain locked within the cranial cavity. And there is the *functional brain,* which is the brainlike tissues and chemicals that serve brainlike functions and may or may not be located within the skull. In other words, there is brain outside the brain—substances functioning in many respects like brain tissue but distant from the bony cranium. The new understanding of the role of the blood makes it appear as if the brain has broken loose from its moorings and is coursing throughout the body wherever blood flows. From this perspective it seems that mind, to the extent that it is connected with the brain, permeates every nook and cranny of the body, for there is little tissue in the body that is bloodless.

These new insights provide another window on nonlocal mind. Here we do not mean mind that is spread through time or that the mind is spread beyond the body, as we saw in our discussions of prayer and psychic healing. Here we are looking at a more conservative meaning of nonlocal mind—mind not localized to the brain, mind not confined to a specific space within the head. Perhaps this step in our thinking could be an intermediate one to fully understanding nonlocality. First, get the mind outside the brain and into the rest of the body, and second, beyond the body altogether. This second step may seems less traumatic if we understand the significance of the first: that the mind, even if we concede that it works through the brain, cannot be fixed within the skull; that it is suffused through the body, coursing wherever our blood flows. We must begin to think about the distant brain, the ectopic brain, the brain-at-large, the blood-born brain, the brain that oozes red from cuts and abrasions when the body is traumatized, the brain that is visible in microscopic and chemical analyses of the blood, the brain that flows through capillaries and venules and arteries from the innermost core of the body to the outermost elements of the skin. This distant brain is capable of affecting the function of the traditional, anatomic brain—*so* capable, in fact, that we simply are not justified in thinking of the brain as something that exists only above the shoulders. We must consider that anywhere there is blood, there is brain also.

If the brain has slipped its moorings, what are we to say about the location of the mind? We even speak of ''losing'' our mind, as if it were a bad thing for it to wander off and not stay put. The great anthropologist Margaret Mead was one of the many who do not share this prejudice that the mind is localized to a specific point in space. The contemporary explorer of the psyche, Jean Houston, relates an incident in which she once asked Mead where her consciousness was located. ''Why,'' the great woman replied, motioning to her entire body, ''it's all over!'' Mead was as good a listener as Hermann Hesse: She could hear the whisperings of her own blood, and they came from everywhere.

"Bad Blood"

Throughout the ages mankind has always known that the proper function of the blood was to insure good health. When things have gone wrong with the body the blood has been frequently blamed, and it is amazing how many diseases throughout history have been attributed to "bad blood." The venereal diseases are the classical examples—particularly syphilis, which ravaged Europe and Asia in the latter years of the fifteenth century in the form of the great pox (as opposed to smallpox). Long before the cause of syphilis was known it was widely felt that "bad blood" was the cause, although there really was no concrete evidence for this, and specialists could not agree on what might be causing the blood to be bad. Some thought the blood became impure because of ethical or moral transgressions (similar views can still be heard today about the cause of AIDS), and called syphilis the "carnal scourge" or the French pox. Others thought the disease was due to specific skin afflictions, while some were convinced that mercury poisoning was the cause.

It seems that everyone in Europe was horrified of being exposed to the bad blood of others. The pandemic of syphilis began soon after Columbus returned from Haiti in 1493, and also coincided with the invasion of Italy in 1494 by the army of Charles VIII of France. Thus the French were convinced the disease was due to bad Italian blood; the Italians were of the opinion that it was French blood that was tainted; and everyone in Europe could blame the natives of the New World for having exported the bad blood via Columbus' men.

It is utterly fascinating that *each* of the primary schools of thought led to the insistent call for blood tests, even *before* it was known that the spiral bacterium causing the disease existed in the blood. Toward the end of the last century researchers looked with a vengeance to the blood for a cause. The search culminated in 1905 with the discovery by Fritz Schaudinn and Paul Hoffmann of *Treponema pallidum,* the thin, delicate, mobile, and strangely beautiful organism whose tapered structure is spiraled six to fourteen times. Yet even before the bacterium was isolated from serum oozing from syphilitic sores, experiments had already confirmed that syphilitics did indeed have bad blood for other reasons—for example, there was a reduced amount of both water and sodium chloride in it. Others found a disturbed "pattern of nutrition" in the blood. None of these findings proved particularly helpful, however, until the discovery in 1906 of the Wasserman reaction, the blood test that allowed syphilis to be diagnosed for the first time. [1] (The end point of this complex test is whether or not the patient's red blood cells are lysed, or dissolved, in the test tube. A positive test requires the presence in the blood of three substances—an antigen [from the bacterium], antibodies [produced by the body in response to the antigen], and a blood protein called a complement.

How did all the disparate ideas about the cause of syphilis—deranged morals, skin diseases, mercury poisoning, and more—center on the conclusion that the *blood* was the culprit in all of them? [2] Where did these hunches come from? Could the uniform, early "reasoning" about the blood as culprit have been a result of a collective listening to the whisperings of the blood?—the blood actually speaking in its own language, just as we say the brain or the heart speaks to us in its language? The image arises of the "hurt organ"—the blood—crying out to be heard; or the person sensing "blood feelings" just as we speak of "gut feelings." Could this be what the early researchers in syphilis were experiencing that led to the common focus on the blood?

This suggestion is not entirely fanciful. The fact is that we have the necessary elements in place for two-way conversations between certain blood cells and our brain tissue. Both organs make many identical chemicals and contain many identical receptor sites. Why would nature have designed this sort of arrangement unless real communication was meant to take place? We know today that chemical conversations between blood and brain *can* occur; our task is to decipher what is being said—to listen, with Hesse, to the whisperings of our own blood.

Gut Feelings

The White people think the whole body is controlled by the brain. We have a word, umbelini *[the whole intestines]: that is what controls the body. My* umbelini *tells me what is going to happen: have you never experienced it?*

—MONGEZI TISO
Xhosa tribesman
South Africa [3]

When your stomach disputes you,
lie down and pacify it with cool thoughts.

—SATCHEL PAIGE
Famous baseball pitcher

Everyone experiences intuitions and hunches from time to time that can't be rationally defended, but that seem unerringly true and irrefutable—"gut feelings" we sometimes call them. That we refer to them by this term is, on the surface, odd: why not "arm feelings," "leg feelings," or "mouth feelings?" Why the gut?

Candace B. Pert, former Chief of Brain Chemistry of the clinical neuroscience branch at the National Institute of Mental Health, was the codiscoverer of the endorphins (with Solomon Snyder, who was awarded the Nobel Prize for his work). Her findings may revolutionize our beliefs about the origins of our feelings, and our assumption that the mind can be localized to the brain. Pert, a preeminent researcher and a genuinely respected insider in this enormously important area, has discovered that every receptor site that she has been looking for within the brain is also found on monocytes, a type of white blood cell that has a pivotal role in the immune response. In addition, she has found that certain chemicals that affect emotion also control the routing and migration of monocytes. Pert's research team has found that these cells of the immune system not only have receptors for various neuropeptide chemicals that control mood in the brain, they also *make* these substances; and that the entire lining of the intestinal tract, from the esophagus through the large intestine, is lined with cells containing neuropeptides and receptors for them. "It seems entirely possible to me," says Pert, "that the richness of the receptors may be why a lot of people feel their emotions in their gut—why they have 'gut feelings.' " [4]

Who could possibly doubt that we "feel it in the gut"? So widespread is the capacity for "gut feelings" that only recently the drug cimetidine, known as Tagamet, was the largest selling prescription medication in the United States. Tagamet is a "miracle drug" used to treat hyperacidity and ulcers. "Gastritis," "indigestion," "gas," pain, and bleeding are the prices we pay for the physiologic truth that we channel feelings into the gastrointestinal (G-I) tract. These problems are not benign: Each year thousands die of hemorrhage from gastric and duodenal ulcers. But there is more to the G-I tract than the stomach: There is the mouth, the esophagus, the duodenum and small intestine, and the colon as well. Although many causative factors are involved in these illnesses in addition to emotional stress, these organs all bear the brunt of the emotions, as countless sufferers of esophagitis, esophageal spasm, peptic ulcers, duodenitis, and irritable bowel syndrome (spastic colon) will attest. And Pert has discovered that *all* these sites have receptor sites for endorphins and the capacity to manufacture these mind-altering chemicals.

Endorphins are not the only chemicals shared by the brain and the G-I tract. In recent years many peptide hormones, previously thought to be found only in the G-I tract, have also been found in the brain, among which are vasoactive intestinal polypeptide (VIP), cholecystokinin (CCK-8), gastrin, substance P, neurotensin, enkephalins (similar to the well known endorphins), insulin, glucagon, bombesin, secretin, somatostatin, and thyrotropin-releasing hormone (TRH). [5] As the number of hormones found in both places has increased, a few thoughtful researchers have realized that deep philosophical questions are being raised by these discoveries. The fact that the gut and the brain make the same hormones, which are chemical messengers, and that they share receptor sites for these chemicals, provides a way for them to "talk" to each other. But if they are talking to each other, who's in charge—the brain, as we've always supposed, or the gut? Or *is* there a physiological dictator, some "king organ," as we have thought? In view of this shared function, which has decided mental effects, how can we continue to regard the brain as the sole site of the mind? And what about the organs in addition to the G-I tract that *also* share hormones with the brain (such as certain blood cells, as we've seen, and the kidney, which shares the peptide angiotensin II with the brain)? Do they also have a claim as a site for mind? For every organ discovered to share hormones with the brain, these same questions arise. In the wake of these discoveries, we are left with the question: *Where in the body is the mind?*

Our sensations about where mentation and feelings are coming from are notoriously unreliable. A heart attack may cause pain in the left arm, hand, neck, or the jaw, but not in the chest; the pain from a missing limb may not be felt in the body at all, but projected completely outside the body (phantom limb pain). So, where do the mood changes I experience originate? They *feel* as if they are originating in my head, but how can I be sure? How do I know they are not coming from the gut—or the blood or kidney—since a similar kind of "emotional equipment" is located there as well as in the brain?

A Bridge Between All Living Things?

One of the most surprising findings to come out of Pert's research about neurotransmitters, peptide hormones, and receptor sites is how similar these substances and structures are to those found in life forms other than human. A tiny protozoa, called tetrahymena, a primitive, microscopic organism, makes insulin and beta endorphins just like those of humans, and its receptor site for the endorphin molecule is identical with our own. Not only this protozoa, but rats as well, share a common receptor site for endorphins with humans. The implications of this sharing are profound. About these discoveries, Pert states,

> The opiate receptor in my brain and in your brain is, at root, made of the same molecular substance as the tetrahymena. . . . This finding gets to the simplicity and unity of life. It is comparable to the four DNA-based pairs that code for the production of all the proteins, which are the physical substrates of life. We now know that in this physical substrate there are only 60 or so signal molecules, the neuropeptides, that account for the physiological manifestations of emotions —for enlivening emotions, if you will, or perhaps, for flowing energy. The protozoa form of the tetrahymena indicates that the receptor molecules do not become more complex as an organism becomes more complex. The identical molecular components for information flow are conserved throughout evolution. The whole system is simple, elegant, and it may very well be complete. [6]

Human and rat brains, as well as microscopic protozoa, are all tied together functionally with identical molecular components, components that in humans influence our ''enlivening emotions,'' as Pert puts it. Frequently we refer to these emotions as those qualities that ''make us human.'' So what are we to say now? Have we underestimated what other forms of life can feel? Do they have some sort of emotional or mental life, since they possess some of the chemical equipment required for such? What are protozoa and rats doing with receptors for endorphins in the first place? Why would they need receptors for endorphin, which means "the morphine within," if they are not "feeling creatures"? Morphine relieves pain, and endorphins are many times more potent than morphine as a pain reliever. If these organisms cannot feel pain, why did they develop the endorphins and the receptor sites that go with them? Presumably these chemicals serve some purpose; nature does not select them for no reason at all. And even if we allow that animals may feel—which is really not difficult, since we can see them writhe and hear them cry out in pain—what about protozoa? Must we not ascribe feelings to them if we follow our logic? In sum, these findings challenge the somewhat arrogant way we have disregarded the potential for feelings among lower life forms.

Do these findings provide a possibility for communication between species? Could the minds of different species interact, given similar "mental equipment" through which mind, whatever it may be, might work? Could the similar "mental equipment" in other life forms be a means for all organisms to "tune in" to mind? Put another way, if mind is nonlocal to the extent of being outside the body, a possibility suggested in the clinical studies above, could similar brains or brainlike organs, by having a similar structure, "invite" nonlocal mind to manifest through them, just as iron filings might invite a magnetic field to manifest through them? In the next chapter, Creatures Great and Small, we will see how this possibility may actually come to life. We will review studies in experimental psychology which suggest that some sort of nonlocal mental connection indeed seems to be occurring between man and lower life forms.

But first, a modest caveat is in order. The eagerness with which many persons embrace new findings such as these points to a hunger of a spiritual kind. We want to believe that our life and mind may be greater than our personal, perishable, physical body, that something may endure when we die. Most of us feel deeply that we *are* more than the chemistry, anatomy, and physiology of our individual body and brain. We do *not* side with Bertrand Russell when he said, "When I die I shall rot, and nothing of my ego shall remain." We want Russell to be wrong; we sense that he *is* wrong; and these hopes and intuitions are fanned when science tells us something new and expansive about ourselves.

William Butler Yeats once railed against the habit of science of reducing everything in nature to dry, dead formulas by saying that (referring to the formula H_2O) he preferred his water mixed with a little seaweed. Just so, we prefer our brains and bodies mixed with a little mind and spirit. But sometimes we forget that science is *still* talking overwhelmingly about the physical and the chemical when it speaks of "mind," if the term is even used at all. Frequently it is *we* who add the seaweed to the water by adding "mind," "soul," or "spirit" to the latest revelations of science about the brain and body. We forget that in science today, materialism still reigns.

So hungry are we that our inner urge to transcend the physical be validated by the all-seeing eye of science that we make vast errors in our thinking, which Alfred North Whitehead called "misplaced concreteness." An example of this error is in assuming that, since people are known to "fall" in love, that gravity is responsible. Isn't love a thing, and doesn't gravity cause things to fall? The mistaken logic hinges on attributing a false concreteness to love. Similarly, if we assume that science has shown that we share "mind" with tetrahymena, rats, and other humans because it has demonstrated that identical neuropeptides and receptor sites exist in these life forms, we fall into the same trap. Just as love is not a material thing and therefore cannot behave like a falling body, neither is the mind a material thing whose existence can be proved by reference to matter.

Or we commit another type of mistake in our reasoning called "category error," the most famous example of which is that of the linguist Alfred Korzybski, confusing the map with the territory. Or eating the menu instead of the meal—and in our case, of equating mind with body, the mental with the physical, the spiritual with the physiological.

The philosopher Gilbert Ryle, who is famous for referring to the mind as the "ghost in the machine," proposed that the entire confusion of mind with body rests on a gigantic category mistake of Descartes, the seventeenth-century French philosopher. Descartes, said Ryle, "was rather like a bewildered foreign student on a guided tour of Oxford. As the library, the dormitories, the chapel, and so on are pointed out to him in turn, the student keeps asking, 'Yes, but where's the *university?*' " [7] We insist that the mind, like the body, be a thing localizable in space, just as the foreign student demanded that the university be at some specific point. In so doing we confuse two different levels of description and equate the mind with the brain and the body—a category mistake that continues to haunt medicine and biology.

It is crucial to keep these levels of description clear. Otherwise we run the risk of confusing the spiritual and the physical. We will never validate our spiritual intuition that we are more than the material brain and body by scrutinizing our monocytes, lymphocytes, neuropeptides, and receptor sites. These are material structures and they will perish when we die, and the fact that they may be identical to those of tetrahymena or some other species will not forestall this event. Why, then, are these similarities important? Why do we dwell on them? Our relatedness with other living forms provides us something we sorely need: a reverence for the life of all creatures great and small, and an expanded view of our place in nature—not as rulers *over* it, but as participants *in* it. As we shall continually see, however, it is the mind, not the body or its parts, for which a genuine case can be made for nonlocality, and ultimately for immortality and oneness.

As we proceed, let us remember, then, that the physiological, the mental, and the spiritual are not equivalent. We must be careful not to put *spiritual* words into the mouth of *physical* science; and if we choose with Yeats to add a bit of seaweed to our H_2O, we ought to remember that the source of the seaweed is invariably ourselves, not science.

Nonlocal Mind in the Laboratory

> *The unconscious has an extension that can reach anywhere, we have absolutely no means of establishing a definite frontier; as we cannot say where the world ends, so we cannot say where the unconscious ends, or whether it ends anywhere.*
>
> —C.G. JUNG [8]

Our look at the bodymind continues—only now we want to focus not on the connections that exist within the individual, personal self, but on the way our minds may be connected to the minds and bodies of others. And not just other humans, but other forms of life as well.

In his remarkable book, *The Invented Reality,* which explores the emerging school of thought known as constructivism, philosopher Paul Watzlawick presents a series of experiments which show that the reality we perceive as ''out there,'' a reality that simply waits for us to discover it, is a false notion. [9] The picture seems to be that reality is not so much discovered as invented.

On the surface this may seem to be making mountains out of molehills. Everybody is aware today that people differ in their perceptions, even when they are confronted with identical situations. Our past experiences, beliefs, and expectations dramatically shape the ways we see, smell, hear, taste, and feel. Even something as simple as the estimation of the length of a line varies tremendously between individuals. If a person has been raised in poverty as a child, he or she will perceive a coin to be larger than does someone raised in wealth, who will see it as being smaller. The examples that could be given showing that human perceptions vary in response to identical sensory challenges are endless. They no longer cause psychologists to raise an eyebrow, so commonplace are they.

But can perceptions and preexisting beliefs *about* the world actually *change* the world? The belief about a line's length does not really change it, nor does the erroneous estimation of the size of a coin change it. The line and the coin are what they are; beliefs about them do not influence their actual nature. But *could* beliefs actually change the state of the physical world?

In the last three decades findings in experimental psychology have suggested that one's belief about the world may actually change it. This idea is very disturbing to the usual conceptions of the mind, suggesting that mind can actually influence events at a distance—that it can "move matter" and thereby shape the world around us.

Watzlawick brings to light the sobering investigation of psychologist Robert Rosenthal of Harvard University known as the Oak School experiment. [10] This study concerned a school of 650 students and eighteen female teachers. At the beginning of the school year the faculty was told an untruth—that a particular intelligence test, given to all the students, could not only determine their intelligence quotients (IQ), but could also identify that 20 percent of students who would make rapid and superior intellectual progress in the coming school year. After the test had been given, the teachers were given the names of the students who supposedly, on the basis of the test, could be predicted to excel. But in fact the teachers were being "set up," for the names were picked *randomly* from a student list. Thus the difference between this group of students and the rest of the student body existed solely in the teachers' heads. At the end of the school year the same intelligence test was given to the student body. The results? The "gifted" group of students demonstrated *real* above-average increases in their IQ's. In addition, the subjective reports of the teachers rated those students to be more outgoing, friendly, and intellectually curious than their peers.

What happened? There are many ways to explain the outcome of this study. We might say that a self-fulfilling prophecy was induced in the teachers, which they brought about through lavishing, unconsciously, extra attention on the "gifted" students. This gave them a learning advantage that manifested itself in the final test score. After all, teachers love to have gifted students, and thus it is not unreasonable to suspect that the teachers were extraordinarily attentive and empathic to their "special" charges.

But the explanation of the outcome of this experiment, as we shall see, is far from clear. It may *not* rest solely on the known principles of experimental psychology, which assume that minds are separate and isolable to single beings. It may require bolder hypotheses about how the mind works than we have up to now been willing to entertain. In fact, it may prove clumsier to believe that there are *5 billion* minds on our planet that are separate and individual, than to suppose that there is one mind that is nonlocal and which manifests through different persons. *If* mind is nonlocal, *if* something like an extended, universal, or group mind is real—*if* these possibilities exist, it would be entirely reasonable to suppose that the beliefs of the teachers changed the performance of the "special" students by merging with them and exerting the observed effects; for if mind really is nonlocal there would have been only one mind to begin with, to which the minds of both teachers and students belonged.

Prior to carrying out his Oak School experiment, Rosenthal had reported a similar study with rats that was repeated and confirmed by many other scholars. [11] Twelve students in a course in experimental psychology were given lectures in which they were told that either good or bad test achievement could be handed down genetically in rats by selective breeding. Then six of the students were given a group of thirty rats which, they were told, had been bred from superior stock, and were therefore good or fast learners. The other six students were given thirty rats and were told the opposite: Their animals were bred for poor performance on learning tasks. Actually there was no difference genetically in the rats at all. All sixty came from genetically homogenous stock, the kind always used in this type of experiment in experimental psychology. All sixty animals were then trained in the same learning experiment. The rats whose trainers believed them to be superior actually did better from the very outset, and raised their achievement scores far above those of the "unintelligent" animals. At the conclusion of the five-day study the observers were asked to rate their animals subjectively. Those trainers who believed they were working with superior animals judged them so. They reported them to be inquisitive, intelligent, and engaging; they said that they even played with them frequently, touching and petting them often. On the other hand, those students who "knew" they were dealing with *un*intelligent animals expressed negative judgments about their subjects.

Can we explain the outcome of this experiment through the usual sort of "observer bias"? Could the results perhaps be due to the subtle influence of the touching, petting, and interaction of the experimenters and the "gifted" rats? If so, there is nothing to distinguish this study from the usual explanations for the outcome of the Oak School experiment: It's all a matter of the subtle cues, empathy, and preferential learning advantages being instigated by the unwitting teacher or experimenter. There is really no need from this perspective to suggest that the *minds* of the experimenters affected the performance of the rats. After all, students are lovable (some of them!), and laboratory rats often are cuddly creatures, too; empathy and outgoing feelings toward both sets of subjects would in fact be expected. Keep the explanations clean, simple, and economic: Don't violate Occam's razor. According to this hallowed maxim in science, enunciated by the English philosopher William of Occam in the fourteenth century, the best explanation is that which makes use of the fewest assumptions. "Multiplying one's hypotheses" beyond the minimum is forbidden—such as, in these experiments, dragging in "mental" effects and the concept of nonlocal mind.

But let us descend further into the kingdom of living things, this time to the world of worms, which most persons, presumably, do *not* find extraordinarily lovable and cuddly. In 1963 two researchers, Lucian Cordaro and J. R. Ison, reported the results of a provocative study dealing with a primitive earthworm called planaria which, they suggested, are not explainable by the usual interpretations of the ways experimenters and their subjects interact.

Since planaria have a rudimentary brain, it has been felt that they may be capable of learning very simple tasks such as the direction to take upon arriving at the crossbeam of a T-shaped groove experiment. These experiments were quite common in American universities in the 1950s. In the Cordaro and Ison experiment, just as in the Oak School experiment and in the above study with rats, the experimenters were led to believe that they were dealing either with genetically gifted or intelligent worms, or with dull and incapable ones. From this belief grew results similar to those we have already seen in the other two studies: Those earthworms the observers believed to be gifted achieved measurable, superior results in their learning tasks. [12]

Although the subjectivity of the experimenters entered into the study in the form of certain expectations about the performance of the worms, it is highly unlikely that the results can be explained through actual emotional attachments between the experimenters and their subjects. For, as Paul Watzlawick wryly states in his review of this study, "at this primitive stage of development [of these worms, there is] little room for emotional attachment." [13] And in all three of these experiments—with children, rats, and worms—it is important to note that we are dealing not merely with subjective endpoints but with measured outcomes. In each study the results could be numerically stated, and the numbers varied in the same direction as the mental predictions and expectations of the teachers or experimenters.

Could subtle cues still be working in ways we can't explain, ways that are invisible but that shape the behavior of the subjects, even between humans and worms? Of course there is *some* interaction going on, and it is only a question of what kind. Wherever empathic effects enter in, or wherever there is positive emotional exchange between experimenter and subject, we can expect these factors to make a difference in performance. But it simply strains one's credulity to believe that the cues in all three of the above experiments could be completely explained on the usual basis—interactions based solely on verbal, tactile, or visual exchanges, or by some other physically mediated interaction. The problem is that these sorts of interactions do not always seem to be present, especially when the life form involved evokes not empathy and warmth from the experimenter, but negative feelings, which is usually the case with worms.

Whatever explanation we choose, it is obvious that we shape our picture of reality, and even the safeguards of science cannot always prevent our role as participators. Watzlawick poses the dilemma arising from these strange studies, and our typical reaction to them:

> For the very reason that these experiments undermine our basic concepts, it is all too easy to shrug them off and return to the comfortable certainty of our accustomed routines. That, for instance, test psychologists ignore these extremely disturbing results and continue to test people and animals with unmitgated tenacity and scientific ''objectivity'' is only a small example of the determination with which we defend ourselves when our world view is being threatened. [14]

Alternatively, the results of the above experiments could be explained by saying that a psychokinetic event is taking place in which an individual mind is ''moving matter'' in ways we simply do not understand, thus ''pushing'' the behavior of the students, rats, and worms in a particular direction. But such a mind would be nonlocal, as its effects would clearly be outside the brain. And if it is nonlocal in space and genuinely unrestricted to the individual, it seems possible that it would merge with other similar minds, forming the One Mind. Psychokinesis, then, would seem to be a special manifestation of the nonlocal, One Mind, and not fundamental in itself.

No doubt there *were* subtle interactions that shaped the performance of the students, rats, and worms in the above experiments, and that some of these interactions were physical in nature—the cues generally used to explain "experimenter bias." The *most* subtle interaction, however, may lie in what Watzlawick calls in passing "human communication." The proposal being put forward in this book is that this "communication" will prove to be entirely different than current theories allow. The known facts about human communication, and communication between humans and other species, will likely be understandable only through the concept of nonlocal mind—the *One Mind*—an enveloping mind pool, a universal consciousness that takes in the minds of all teachers, experimenters, students, rats, and worms alike. This mind is not confined to single observers, although it is manifested through them, creating within them the illusion of a separate self and the sensation of an ego that possesses a separate mind.

4

Creatures Great and Small

Let's sit down here, all of us, on the open prairie. . . . no blankets to sit on, but feel the ground with our bodies, the earth, the yielding shrubs. Let's have the grass for a mattress, experiencing its sharpness and its softness. Let us become like stones, plants, and trees. Let us be animals, think and feel like animals.

—LAME DEER
Sioux medicine man [1]

Then she sang again and went out of the tepee; and as the people watched her going, suddenly it was a white bison galloping away and snorting, and soon it was gone.

—BLACK ELK
Sioux medicine man [2]

We now take a final look at the way nonlocal mind manifests in the large-scale world of living things before venturing into the world of science in Part II. While we have seen evidence of possible connections between humans and animals in the psychology experiments above, let's look at this possibility outside the laboratory.

If nonlocal mind is possessed by living things below us, then it is extremely likely that our own consciousness contains the quality of nonlocality also. For we believe that our minds can "do" what lower minds are capable of, and more. Indeed, that is why we call our own intelligence "higher."

There is good evidence, from both a historical and a modern scientific perspective, that something resembling mental communication goes on between humans and animals; that this communication can be shockingly profound; that it cannot be explained by known physical interchanges; and that it is at times decidedly nonlocal in nature. For many ancient cultures, it would have been unthinkable that we do *not* share consciousness with other life forms. As an example, consider the following comment by Tatangi Mani, or Walking Buffalo, an American Indian who lived from 1871 to 1967:

> Did you know that trees talk? Well they do. They talk to each other, and they'll talk to you if you listen. . . . I have learned a lot from trees: sometimes about the weather, sometimes about animals, sometimes about the Great Spirit. [3]

Here we have the picture of a world alive and buzzing with intercommunication between man, plants, and animals, a world in which all of nature is one big conversation. Humans participated in the talk for several reasons, some practical, such as knowing the weather, and some spiritual, such as knowing about the Great Spirit. This is a picture of nonlocal mind in action—mind loose in the world, mind ranging beyond the brain and body to communicate in ways for which we simply have no known physical explanations.

For those so-called primitive cultures the entire world was alive. It was *enchanted, actually* alive for them, and all forms of matter possessed mind. To suggest otherwise would have been considered a blasphemy against nature. Moreover, our modern idea of local mind would have been considered a downright dangerous belief because it represented the kind of hubris that the gods would never leave unpunished.

Sharing Mind with Animals: Shamanism

Although all members of primal societies were generally aware of an enchanted, conscious world, they were not equally skilled in traversing it. The master in primal cultures in understanding and accessing this living reality was the shaman, and it is from the study of shamanism that we can learn much about the interactions of the mind of men and animals. ''Shaman'' comes to us from the Russian, from the Tungusic *saman*. In the vast area of Central and North Asia, whence the word originates, the shamanic role lay at the heart of the magicoreligious life of all tribes. But not only there; the shaman existed in cultures from the Arctic to the countries near the southern pole; from North America to Europe and Southern Asia; in Indonesia, Oceania, and elsewhere. He was ubiquitous because the needs he served were universal: the need to know the spirit world and how it affected the affairs of men. His role was broad in all cultures. He was healer, manipulator of sacred rites, prognosticator of the hunt and the harvest, intercessor for lost souls—in general a specialist in bridging the mundane and the sacred worlds for his people.

Ethnologists have studied the role of the shaman in cultures all over the earth, and today there can be little doubt that shamans everywhere believed that a kind of collective consciousness bound them together with the animal kingdom. So intimate was the sharing of mind with the animals that shamans believed it possible to actually *become* an animal. According to the brilliant scholar of shamanism, Mircea Eliade:

> Relations between the shaman (and, indeed, "primitive" man in general) and animals are spiritual in nature and of a mystical intensity that a modern, desacralized mentality finds it difficult to imagine. *For primitive man, donning the skin of an animal was becoming that animal*. . . . even today, shamans believe that they can change themselves into animals. . . . We have reason to believe that this magical transformation resulted in a "going out of the self." (emphasis added) [4]

So at once we notice a profound difference in the way the shaman viewed animals and the way we think about them today. So personal was his relationship with the animal kingdom that he would never use the word "animalistic" in the pejorative sense in which it is commonly used today, for animals were not seen as lowly creatures, but were in fact closer to God, the ultimate source of power, than we. There was much to be gained, therefore, in becoming an animal. Eliade:

> The animal with which the shaman identified himself was . . . a mythical animal, the Ancestor or the Demiurge. By becoming this mythical animal, man became something far greater and stronger than himself. We are justified in supposing that this projection into a mythical being, the center at once of the existence and renewal of the universe, induced the euphoric experience that, before ending in ecstasy, showed the shaman his power and brought him into communion with cosmic life. [5]

Transforming oneself into animals was not limited to shamans. A martial technique existed among the Indo-European peoples, parallels to which have also been found in extra-European cultures, in which the warrior transformed himself into a beast of prey, taking on the ferocity and savagery of the animal in question. These "wild beast warriors" were called the *berserkir* (to which our word "berserk" is related), and shows that communication and transformation between animals and man extended far beyond the domain of the shaman. Another expression can be found in the hunting rites of the paleo-Siberian peoples, in which the transformation into the animal of prey afforded the hunter a great advantage, since becoming the hunted animal helped keep one from going hungry. [6]

There are reports of the shaman who, taking on the consciousness of the heron, disappears suddenly from the sight of those around him, leaving only the sound of flapping wings; or the shaman who, becoming a bear, leaves only bear tracks instead of his own at a certain place on the trail. Do these transformations really occur? No doubt trickery is involved in many of these reports (it has been documented, and even admitted by some shamans), but not in all. But it is extremely unlikely that any tradition could exist across scores of millenia to the present day if its sole support was trickery, deception, and lies.

Perhaps it is not important to give a true-or-false answer to what "really" goes on in shamanism, but to accord a great respect to this cultural tradition that has existed since the dawn of human history. Eliade himself implies there is something ineffable about these practices—something awesome, something spiritual, on which the modern skeptical mind might do well not to trespass. We are, after all, no longer familiar with the nonordinary states of consciousness that are well known to the shaman, a fact that surely limits any observation we might make about it. Shamanism will never yield all its secrets to the intruding observer, especially to one who rejects as pathological the ecstatic psychological domain that is central to its operation. The objective descriptions of the anthropologist can never fully capture the essence of shamanism, just as a scientist cannot adequately describe Michelangelo's *David* by merely giving an account of the molecular structure of the marble from which it is sculpted.

Sharing of animal consciousness brought with it certain skills that were hard to implement in the ordinary, mundane world. In day-to-day consciousness man is bound to the laws of cause and effect. He is tied to linear time with its rigid divisions of past, present, and future. By and large, with occasional exceptions such as in precognitive dream states, he is able to live only in the present, although he strains mightily to get a glimpse of things to come. But by becoming or sharing the consciousness of an animal, the shaman believes he can transcend time and space and the laws of cause and effect that are part of them. For the shaman, animals seem to be part of a mythical time and a paradisical existence before the world took on its present character. Thus he believes that, as an animal, seeing into the future is not problematic, nor is communing with the past. In this state, certain things become possible that cannot be done in profane existence, and he can put this ability to good use. He can find tribesmen who are lost on the tundra, the mountain, the desert. He can locate strayed horses. He can find water. He can foretell the harvest. He can predict the hunt. He knows where game is. He knows who is going to die and who will live. He knows the outcome of battles, and where and when it is best to fight. He can predict the weather and when to plant or harvest. All of this because the cosmology, the world experience, of animal consciousness is different: It is not tied to the temporal and spatial constraints of everyday life. It is nonlocal in space and time. Whether or not animals themselves, in "animal consciousness," can really engage in such perceptions is quite beside the point. It is the shaman's *belief* about the nature of animal consciousness that seems crucial, and which enables these heightened powers of perception to flower.

Each time a shaman succeeds in sharing in the animal mode of being, he in a sense reestablishes the situation that existed *in illo tempore,* in mythical times, when the divorce between man and the animal world had not yet occurred. [7] This gave him a fresh start on things, a way of annulling the drifting, decaying effects of linear, profane time. By becoming an animal the world was made vital and new.

In the nonlocal, collective consciousness that wrapped man and animal together, it was not always man who took the initiative in actualizing it. Sometimes the first overture was made by the animal. This is most obvious in the *call* of the shaman—the beckoning to him to become such a master—and in his initiation. Frequently during dreams or vision quests the shaman-to-be confronts beasts who may devour him or torture him in the most hideous ways imaginable. In the end he is generally reconstituted, and not uncommonly the initial devouring beast becomes his tutelary guide or protective animal. In the tradition of the Buryat shamans the tutelary animal is called the *khubilgan,* a term that can be interpreted as "metamorphosis" (from *khubilku,* "to change oneself," "to take on another form"). Thus the tutelary animal not only enables the shaman to transform himself; it is in a sense his "double," his alter ego. This alter ego is one of the "souls" of the shaman, his "life soul." [8] If his alter ego is killed in a fight, the shaman very soon thereafter dies. [9]

Shamanic phenomena have often been ascribed to neurophysiological abnormalities. But as S. F. Nadel stated in his study of Sudanese tribes, "No shaman is, in everyday life, an 'abnormal' individual, a neurotic, or a paranoic; if he were, he would be classed as a lunatic, not respected as a priest. Nor finally can shamanism be correlated with incipient or latent abnormality; I recorded no case of a shaman whose professional hysteria deteriorated into serious mental disorders. . . . [One cannot, then, say that] shamanism . . . absorbs mental abnormality at large [or that it] rests on uncommonly widespread psychopathic predisposition; it certainly cannot be explained merely as a cultural mechanism designed either to achieve the former or to exploit the latter." [10]

Indeed, the shaman generally seems to have a more than normal nervous constitution, for he must be capable of extraordinary concentration. And by any standard, his intellectual skills far surpass those of his fellows. For instance, the poetic vocabulary of a Yakut shaman contains 12,000 words, whereas the commonly used language has only 4,000. Intellectually the shaman lies above the milieu, frequently the guardian of a rich oral heroic literature, as well as being the singer, poet, musician, diviner, priest, doctor, and the general protector of legends several centuries old. [11]

And the shaman's professional duties are so demanding that few can match his physical prowess. He must be able to endure the solitude and danger of the wilds; he must be able to withstand the rigors of the fast and the hours of dancing and chanting. Thus a supple, strong body and an unbounded energy are required to supplement his keen intelligence. As one account has it, "Mytchyll, a Yakut shaman . . . , though an old man, during a performance outdid the youngest by the height of his leaps and the energy of his gestures. . . . He became agitated, bubbled over with intelligence and vitality. He gashed himself with a knife, swallowed sticks, ate burning coals. . . ." [12]

The shaman also excels politically and socially. "For the Yakut, the perfect shaman . . . must be serious, possess tact, be able to convince his neighbors; above all, he must not be presumptuous, proud, ill-tempered. One must feel an inner force in him that does not offend yet is conscious of its power." [13] The best shaman, then, would be a hybrid of a great athlete, an unstinting intellectual capable of synthesizing great stores of knowledge, and the noblest statesman on earth.

Shamans as a group are surely among the most talented and brightest humans who have walked the earth. The belief in a nonlocal, shared consciousness with animals comes to us from an elite class of human beings. This fact does not mean that they were not wrong. Maybe they were collectively deluded or their notion that the entire world of animals, plants, and things was enchanted and alive with mind was a holdover from the days before man's consciousness had totally dissociated from its identification and fusion with the world around him; when his ego and concept of self was so fragile that he did not have a developed sense of "I." Or perhaps, having developed a sense of self, the shaman in every case is merely regressing to a coveted time of paradisiacal innocence.

Sharing Mind with Animals: Modern Evidence

If we wanted to look to the modern world instead of the world of the shaman, where would we go for evidence of a shared mind with animals? The evidence may be everywhere. For instance, it would be difficult to convince millions of pet owners that they do not on occasion "communicate" with their dogs, cats, birds, or even their goldfish. Everyone has felt at one time or another on the same "wavelength" with an animal. This feeling is particularly pronounced in some vegetarians who feel so strongly connected to animal life that they cannot bring themselves to eat animal flesh. It is impossible to explain the commitment to vegetarianism in some persons on the theory that it is purely an intellectual decision. Something more is involved—such as, perhaps, a genuine sensing of the shared consciousness between humanity and the animals. In a similar vein the growing movement against animal experimentation may represent the same felt connection. In these three instances—pet ownership, vegetarianism, and the reaction against animal experimentation—we may be witnessing vestiges of the time when humanity actually felt a sharing of consciousness with the animal kingdom.

Sometimes the animals may return the favor, behaving in ways suggesting that *they* share consciousness with *us*. Could this be the fuel behind the euphemism that the dog is "man's best friend"? Why should he want to be? For table scraps only? Maybe, although I am not convinced. What about the legendary stories of animals coming to the aid of humans in distress, such as dolphins saving drowning sailors, or dogs or horses leading lost persons to safety? The possibility cannot be dismissed out of hand that they are responding to *their* feelings of a shared consciousness with *us*.

The biologist Lyall Watson has described numerous extraordinary instances wherein persons seem to somehow communicate with other species. [14] In one Polynesian village a man awakens from a dream and goes down to the lagoon, which is some distance away and completely out of sight, to meet the sea turtles or the dolphins because he *knows* that they are there at that very moment, waiting for him. Or he may lean over the rail of the fishing boat, thrust his head under the water, and "know" the direction in which the school of fish lie, how far off they are, and what particular kind of fish are in the school. It may well be that certain physical senses are being utilized in these and similar instances. But we have no known theories to account for them as biology is currently understood.

One person who is convinced that one mind envelopes humans and animals alike is J. Allen Boone. Boone is the eloquent author of *Kinship With All Life,* [15] a movie producer, and the ex-head of RKO movie studios. Boone's experience with Strongheart, an international champion German shepherd and movie celebrity, left a deep impression on him and changed forever his ideas of mind.

When Boone gained possession of this famous dog, he was advised to treat him as though he were an intelligent human being. He was instructed to say nothing to him verbally he did not feel in his heart, and was told to read something worthwhile to him each day. But shortly following Strongheart's arrival at Boone's home, conflicts of life-style arose and Boone complained to the dog. Strongheart listened, then began to pantomime the reasons for his behavior. Boone describes their interchange:

> I had spoken to Strongheart in my kind of language, a language of thoughts and feelings incased in human

> sound symbols. He had actually been able to receive and understand what I had said. Then he answered me in a kind of language, a language made up of simple sounds and pantomime which he obviously felt I could follow without too much difficulty. Strongheart had understood me perfectly, and then with his keen and penetrating dog wisdom he had made it possible for me to understand him, too. [16]

Sheer fantasy? Was Boone merely projecting his own feelings into the situation? He is certain this was not the case:

> For the first time, I was actually conscious of being in rational correspondence with an animal. . . . I had been privileged to watch an animal, acting upon its own initiative, put into expression qualities of independent thinking . . . clear reasoning . . . good judgement . . . foresight . . . prudence . . . common sense. I had been taught to believe these qualities belonged more or less exclusively to the members of the human species, or rather to the "educated members" of our species. And here was a dog overflowing with them! [17]

Boone's experience so far is perhaps not unique. Millions of pet lovers are certain their pets are "part human." But he went further and elaborated in his writing an ethic and a philosophy of universal mind that connects human beings, animals, and all living things—including, it turns out, *flies*.

Following his remarkable experience with the German shepherd Strongheart, Boone made friends with a common housefly he named Freddie. Every morning Freddie greeted Boone by landing punctually on his shaving mirror at seven o'clock. The fly learned to come when called by his name, and Boone would play a game with Freddie. He would invite him to climb aboard his finger, then he would stroke his wings gently. In a variation, he would toss the fly in the air and catch him again on his fingertip.

Recalling what he had learned from his experience with Strongheart the German shepherd, Boone's experience with Freddie the fly led him to formulate the following beliefs:

> (1) That inherently Freddie the fly and I as living beings were inseparable parts of an interrelated, interfunctioning and all-including Totality. (2) That neither he nor I were originating causes for anything, but instead were individual living expressions of a universal divine Cause or Mind that was ever speaking and living itself through each of us and through everything else. [18]

Boone's credo wasn't just an intellectual construct, it was an experiential program. He *lived* his belief that he was included in a totality that enveloped all beings. He maintained that there was much to be learned by "silently talking across to [Freddie]. Not as to 'a fly' with all the limiting and condemning things that we humans usually fasten on flies, but as to an intelligent fellow being." [19]

It is unlikely that skeptics about nonlocal mind would count as "evidence" communication between humans and dogs, let alone flies. For the doubters, mind remains a sole human possession encased in brains and bodies, and does not stray beyond our species.

In a remarkable series of experiments initiated by Pierre Duval and Evelyn Montredon in France and continued in America by Walter Levy, gerbils and hamsters were placed in a small cage that was partitioned into two halves. The wall dividing the cage was low enough so the animals could jump over it without much difficulty. Once every minute or so a five-second shock was applied to the floor of one of the halves, the sequence of shocks being determined by a random-number generator. The motivation of the animals was presumably to avoid pain by jumping to the side of the box where the floor was not being shocked. But how could they anticipate which side was going to be shocked next? They could not decipher a pattern to the shocks, because there was none: The shocks were delivered randomly. Still, they seemed to know; for when the positions of the animals were recorded electronically and the data analyzed by computer, the gerbils and hamsters somehow managed to be in the right place at the right time to avoid the shocks, at a rate that was far higher than predicted by chance. [20] This study suggests the possibility, at least, that the gerbil and hamster mind, if we may call it that, is not confined to the present, but can range into the future, survey it, and provide meaningful information to the animal.

But could this be a case of telepathy that, according to Boone's model, could occur if the minds of all living creatures are united in a universal mind pool? If his model were true the experimenters could telepathically provide the subjects with the information they needed to avoid the noxious shocks. But not in this experiment; for the experimenters *also* could not know which side of the box was to be shocked next, since the sequence was determined entirely by a random-number generator.

In another study in which animals seemed to be able to know the future, experimenters, C.E.M. Bestall and James Craig placed rats and mice in a maze that provided them with two choices. The animals were able to predict which of the two choices was the correct one and turn in the right direction, when the correct choice was only *later* assigned by a random procedure. [21]

Studies such as these suggest that the minds of animals can violate the "now." But they can also violate spatial constraints and function beyond the "here."

In one particular experiment, a mother dog and one of her puppies were taught to cower when a rolled piece of newspaper was raised. Then the dogs were placed in separate rooms and the puppy was threatened by the rolled newspaper. At the very moment he cowered, the mother did also. [22]

In another study a boxer was attached to an electrocardiogram (EKG) machine in a soundproof room, while his female owner was separated from him in a distant room. Then without warning to her a stranger burst suddenly into the room and began shouting abusively and threatening her with violence. At that moment the EKG recorded a violent increase in her dog's heartbeat in the soundproof, separate room. [23]

But by far the most remarkable evidence that the mind of animals may violate spatial confinement are the innumerable "lost animal" cases. These are extremely common, and almost everyone has heard of an example or two. They seem so much more intriguing than the artificial experiments set up in laboratories. But this is not surprising. If the ability of the animal mind to transcend time and space arose through the evolutionary process, it most certainly did so in the wild and not in the laboratory. Thus it is not surprising that the strongest demonstration of nonlocal mind in animals, if it exists, would occur in natural settings.

As author Bill Schul points out in his captivating collection of animal stories, *The Psychic Power of Animals,* [24] it really isn't surprising if a creature such as a homing pigeon finds its way home if it is carried in a closed box for miles and then released. It is believed that homing creatures rely on many sorts of cues, such as the positions of sun and stars and geomagnetic fluxes. Yet these explanations don't tell us how animals find unknown destinations or places they've never been.

But great caution is required in evaluating "returning animal" stories. There are a lot of lookalike animals, and the family dog that returns may not be the one that was lost. Yet stringent precautions can be taken to avoid deception, such as in the fifty-four "returning animal" cases collected by Dr. J. B. Rhine and Sara Feather of the Parapsychology Laboratory at Duke University. Rhine termed this ability in animals "psi trailing" (psi is a Greek letter used to designate a variety of nonlocal parapsychological phenomena). To qualify as a reportable case, the source of information had to be deemed reliable;

there had to be a specific characteristic of the animal, such as a name tag or identifiable scar; the case had to make good general sense and had to be internally consistent; and there had to be adequate supporting data, such as independent corroboration by other observers, or availability of the animal for inspection by the researchers. Fifty of the cases involved dogs or cats, and four involved birds. Almost half the cases involved distances of more than thirty miles. [25]

A typical "psi trailing" case is that of Bobbie, a young female collie. She was traveling with her family from Ohio to their new home in Oregon, to which Bobbie had never been. During a family stop in Indiana, Bobbie wandered off and could not be found. Finally the family abandoned their search and proceeded on their journey. Almost three months later Bobbie appeared at the doorstep of the new home in Oregon. There was no mistaking her: She still had her name collar, in addition to several identifying marks and scars.

When the story hit the papers, a man named Charles Alexander decided to investigate Bobbie's odyssey. He placed ads in newspapers from Indiana to Oregon and turned up a number of people who had cared for a stray collie who matched Bobbie's description during that time period. "When the trail was mapped out it was found that Bobbie picked a very reasonable route with few detours." [26]

A remarkable collection of returning animal cases was assembled by the great Texas naturalist, Roy Bedichek. Bedichek was an eloquent writer (one of the best Texas ever produced, according to another Texan, Pulitzer Prize-winner Larry McMurtry) and an encyclopedic authority on the fauna and flora of the state. During his life Bedichek combed every part of Texas and recorded his observations in arresting prose. One of his favorite haunts was the Aransas National Wildlife Refuge located along the Texas Gulf coast between Galveston and Corpus Christi. This wild land, which serves as wintering grounds for the endangered whooping crane, comprises 47,000 acres of marsh, salt- and freshwater lakes, brushy sand dunes, coastal prairie, and savanna. The Refuge had proved particularly hospitable for deer, which are found in such numbers that they are trapped and relocated elsewhere in the state as a means of population control. Following capture a deer is tagged by having a piece of metal bradded in the ear, and is then transported sometimes hundreds of miles from the Refuge.

Deer have a homing urge that is quite strong, and over the years many have returned to the Refuge and have been trapped again. From the point of release they somehow thread their way for hundreds of miles among farms, ranches, towns, and cities, braving human and canine enemies alike. Bedichek relates several cases dating in the 1940s, a period in which he spent much time in the Refuge. In the spring of 1942 a doe released near Goliad, ninety miles due west of the Refuge, returned to be trapped again in twenty-one days. In the spring of 1944 a six-point buck was trapped and transported four hundred miles west to Sheffield, Texas. In the fall of 1945 he was again caught in a trap on the Refuge. This pathfinding ability occurs early in life. A fawn was caught in the autumn of 1946 and was driven one night to San Angelo, Texas, 390 miles to the west of the Refuge, and from there to Nolan County, 77 miles to the north. Only two weeks later the same fawn was caught in another trap just 125 yards from the site of the original capture in the Refuge.

These cases are particularly compelling because the possibility of mistaken identity is eliminated: All animals are tagged before release. Could the animals be wandering aimlessly in all directions, a few finally meandering back to the Refuge by sheer chance? Bedichek thinks not; there is only one documented case of a tagged deer wandering westward *away* from the Refuge following release.

Bedichek, a sensitive, lifelong observer of animal behavior, is awed by these instances. "Something draws these wanderers back," he says. "One observer may call it 'Providence,' another 'instinct,' but whatever it may be called, the mystery remains. . . . This is mystery, deep mystery!" [27]

Yet the depth of the mystery could be diminished by viewing the mind of the returning animal nonlocally. After all, the mystery depends largely on the assumption that the animal mind is confined to its brain, and that it therefore cannot know "at a distance." Alternatively, if the animal mind were not confined to points in space (its own brain and body) or time (the present moment), it would be free to scan space and time and gain knowledge of "the way home" over trails it never traveled before. Another possible explanation for how animals return over great distances may lie in shamanism and telepathy. If the record from shamanism is correct and meaningful interspecies communication does exist between humans and animals, then human knowledge of "the way back" could conceivably be shared with the animal mind. Thus in the above case of Bobbie, the lost female collie, the information of how to travel from her home in Ohio to the new residence in Oregon, where she had never been, could have been conveyed to her by her owners, who did know the way. And similarly, Bedichek's returning deer may have gained their knowledge from their abductors.

Dr. Robert L. Morris, coordinator of research at the Psychical Research Foundation in Durham, North Carolina, believes the evidence for psi in animals is important. He states, "The finding of considerable evidence for psi in animals already has one important implication. It tells us that psi is apparently not of recent evolution . . ." [28] If Morris is correct—if nonlocal mind did manifest long ago in animals during their evolutionary ascent—we might expect, according to the tenets of evolutionary biology, that it would appear in humans as well. Have we not designated ourselves *Homo sapiens,* the sapient man, the creature who is superior mentally to all others?

Evolution provides another perspective on nonlocal mind. It is always interesting to ask whether or not a phenomenon whose existence is being questioned has any survival value for the creature suspected of possessing it. If so, this may count as evidence for its existence, since those traits that favor survival tend to be perpetuated genetically. A mind that is shared with animals would seem to have survival value for humans; and studies suggest, indeed, that it *is* good for our health.

In one study it was shown that ownership of a pet was the most important factor in predicting a patient's clinical course following a heart attack—more important than any of the classical coronary risk factors such as smoking, high blood pressure, high cholesterol levels in the blood, or diabetes mellitus. [29]

Or take the case reported by Dr. Gustav Eckstein in his book *Everyday Miracle*. He describes a small spitz dog who doubled as a night nurse for his mistress, who was a diabetic. Each night the little dog would curl up in the angle of the woman's arm. He would awaken immediately if her breathing pattern changed, which is one of the telltale signs of ketoacidosis, one of the most dreaded complications of diabetes. [30]

A growing body of knowledge has accumulated in recent years showing that close connections with animals is valuable for human health. This new area, which bridges traditional and veterinary medicine, is frequently referred to as "pet therapy." [31]

Most instances of human-animal communication can be explained by ordinary sensory cues, and it is unnecessary to introduce the concept of shared mind. But not all of them. If it makes good biological sense that a nonlocal, psychological communion might have evolved between humans and animals as an asset to survival, then the stories of returning animals, precognitive gerbils, and talented flies may be more than amusing parlor tales. They may be indicators that nature in its wisdom would, in fact, have designed a mind that envelops all creatures great and small.

Saint Francis: Patron Saint of Nonlocal Mind

The idea of nonlocal mind implies that our individual mind is part of something larger, something we cannot claim as our own private possession. Taking this idea to heart requires a certain humility—not humility as some existential sweetness that is still only a coverup for the ever-ingenious ego, but humility as the genuine article. It is humility that allows us to know deeply that consciousness is not the sole possession of an ego; that it is shared by not only other persons, but perhaps by other living things as well. It is humility that allows us to take seriously the possibility that we may be on a similar footing with all the rest of God's creatures.

This possibility has been an immediate, felt experience for many cultures, such as the Native American. The Sioux Chief, Luther Standing Bear, expressed this connection:

> Kinship with all creatures of the earth, sky and water was a real and active principle. For the animal and bird world there existed a brotherly feeling that kept the Lakota [Sioux] safe among them and so close did some of the Lakotas come to their feathered and furred friends that in true brotherhood they spoke a common tongue. [32]

The greatest personality in the Christian tradition who embodied this trait most vividly was Saint Francis of Assisi, whom most people regard as a meek, gentle man who loved all animals and who would not harm a flea. But this view hardly fits the picture given of him by an expert on his character, the medieval and Renaissance historian, Lynn White, Jr. Professor White has called St. Francis "the greatest radical in Christian history since Christ," because he championed a pure democracy among all God's creatures. This view—that other living things have an equal status in the world as ourselves—was and still is at odds with the traditional Western view of nature. Yet it is becoming necessary to take it seriously, if we wish to preserve life on Earth, including our own. Thus it is that Professor White has proposed St. Francis as the patron saint for ecologists because of his deep understanding and respect for the Earth and its creatures. [33]

White believes that "more science and more technology are not going to get us out of the present ecologic crisis until we find a new religion, or rethink our old one.'' He proposes that the Christian axioms that nature has no reason for existence save to serve man and that man should exercise dominion over it are in large measure responsible for the desecration that we have mercilessly visited on our planet. But White does *not* exhort us to abandon our religious traditions in favor of a more Earth-oriented philosophy such as might be found in the oriental or Native American traditions, for he is dubious of their viability for us. He champions a rethinking and refeeling of *our own* basic, indigenous attitudes toward nature. But not just our intellectual or rational attitudes. He believes that the roots of our ecological problems are largely religious, and that the solutions to them must also embody religious and spiritual elements. This is where St. Francis comes in: The spiritual foundation for a reappraisal of our attitude toward nature can be found, he believes, in the heretical Franciscan belief in ''the spiritual autonomy of all parts of nature.'' [34]

St. Francis, says the legend of the wolf who ravaged the land around Gubbio in the Apennines, *talked* to the wolf. He *reasoned* with him and persuaded him of the error of his ways. The wolf repented, and when he died he was buried on consecrated ground. This Franciscan doctrine of the animal soul was hurriedly stamped out. St. Francis's view was simply too radical, for it rested on ''a unique sort of panpsychism of all things animate and inanimate, designed for the glorification of their transcendent Creator. . . .'' [35]

There is an unmistakable nonlocal flavor to St. Francis's carryings-on with ''the creatures.'' No one makes brothers of ants and sisters of fire, to say nothing of preaching to birds and wolves, unless there is some deeply felt connection, some going beyond the self. Saint Francis's belief in the virtue of humility, which made possible his intimate discourse with the creatures, is a prerequisite for entertaining the possibility of nonlocal mind. Lacking humility, the idea of a genuinely nonlocal quality to consciousness becomes offensive. It is a threat to the ego, the person, the sense of I-ness that has become enshrined as an element of culture in the West. Possessing this humility in abundance, St. Francis was well positioned for the belief in a genuine sharing of consciousness with the creature-world, and all his actions are compatible with this belief.

Perhaps there is a good reason why St. Francis could talk to the creatures; why shamans, since time immemorial, have ''become'' animals; and why lost animals return home, provided someone, such as their owners, knows the way back. The explanation may lie in the unbounded, shared nature of the mind—the eternal, infinite One Mind, the Universal Mind—that is a central theme of this book. Lyall Watson, an astute naturalist, roving observer, and author of many books, puts it this way:

> I think that there may well be a flow of pattern or instruction which crosses species lines and allows even radically different organisms to borrow each other's ideas. . . . As a biologist, I am aware at times—especially when steeped in some natural cycle—of a kind of consciousness that is timeless, unlimited by space or by the confines of my own identity. In this condition, I perceive things very clearly and am able to require information almost by a process of osmosis. I find myself, at these times, with knowledge that comes directly from being part of something very much larger, a sort of global ecology of mind. And the experience of it is literally wonderful. [36]

Watson implies that if we want to know the One Mind—*our* Mind—we may discover it by going, with St. Francis, through nature.

If we find the idea of the unbounded, One Mind foreign, the reason may be that we have gradually lost our connection with the natural world. As a consequence, our world is now gravely imperiled by our mindlessness, our loss of Mind, our lack of sensitivity to the whole. For loss of sanity, loss of Mind, and loss of Earth go hand in hand. And if we wish to preserve our world, we must first find our Mind by recovering our connections with the heavens and the Earth. To learn to prefer once again, with Yeats, a little seaweed in our water; to begin once more to talk, with St. Francis, to the creatures.

Part II

SCIENCE: The Proofs

The idea of a universal Mind or Logos would be, I think, a fairly plausible inference from the present state of scientific theory; at least it is in harmony with it.

—SIR ARTHUR EDDINGTON [1]

When we view ourselves in space and time, our consciousnesses are obviously the separate individuals of a particle-picture, but when we pass beyond space and time, they may perhaps form ingredients of a single continuous stream of life. As it is with light and electricity, so it may be with life; the phenomena may be individuals carrying on separate existences in space and time, while in the deeper reality beyond space and time we may all be members of one body.

—SIR JAMES JEANS [2]

5

The Immortal, One Mind: Schrödinger, Gödel, Einstein

Mind is by its very nature a singulare tantum. *I should say: the overall number of minds is just one.*

—ERWIN SCHRÖDINGER [3]

Now we leave behind the experiences of patients, prayer, healing, and the coronary care unit to examine the evidence eminent scientists have amassed about nonlocal mind. After all, scientists, rather than poets and mystics, are those who we find most convincing today. Some people believe science has already made a dismal, final pronouncement on the mind: That it evolved through the tortuous paths of the evolutionary process, somehow emerging as an expression of matter, a property of the brain, appearing at a certain stage of biological complexity. When the brain dies, mind dies, and that is the end of the matter. This is a picture of a purely local process, sanctioned by modern science.

But not all of science agrees. Scientific theories have always had their dissenters. When it comes to theories of mind, the views of some of the greatest scientists of our century show that nonlocal mind is not a theory held only by the poetic, mystical, deviant, or softheaded in our culture.

In this section we want to explore the views of three towering figures in twentieth-century science who believed in unconventional views of the mind and the person—physicist Erwin Schrödinger, mathematician-logician Kurt Gödel, and Albert Einstein.

Of the three, Schrödinger's position is most well-developed in his writings. He believed in a collective consciousness or group mind for all mankind, which he called the One Mind. He believed this Mind was eternal, that it was indestructible by Time. Schrödinger was well aware that this idea was contained in ancient wisdom, but he also believed that its outlines could be drawn from modern science as well.

Kurt Gödel, one of the undisputed mathematical geniuses of this century, also believed in the One Mind of which each person is a part. And the most famous physicist of all time, Einstein, had a complex view of human beings and human mind. His idea of the individual is extremely unconventional and to a large extent nonlocal. As we shall see, Einstein could not even make a case for individual, independent will; he saw will as diffused through all persons and events throughout time.

The noteworthy views of these outstanding men are an antidote to the mechanistic, materialistic view that prevails today in conventional science, which tells us that the mind is solely a function of the local, mortal brain and body. These three scientists tell us that another view is possible, one they believed was not inconsistent with other scientific evidence. It is a view of man beyond the individual and the body, of man beyond space and time.

Erwin Schrödinger

> *To divide or multiply consciousness is something meaningless. In all the world, there is no kind of framework within which we can find consciousness in the plural; this is simply something we construct because of the spatio-temporal plurality of individuals, but it is a false construction. . . . The categories of* number, *of* whole *and of* parts *are then simply not applicable to it; the most adequate . . . expression of the situation being this: the self-consciousness[es] of the individual members are numerically identical both with [one an]other and with that Self which they may be said to form at a higher level.*
>
> —ERWIN SCHRÖDINGER [4]

One of the boldest ventures by a scientist into the domain of soul-like, nonlocal mind was taken by one of the great physicists of this century, Erwin Schrödinger. Schrödinger, by any measure, was one of the preeminent architects of modern physical theory: his famous wave equations lie at the foundation of quantum mechanics. But his contributions did not end with physics. Although less well known, he also provided a blueprint for recovering the soul—by showing that the consciousness of humankind forms a unity, and that it is immortal and infinite.

"At the Risk of Making Fools of Ourselves"

In 1958 Schrödinger published his widely acclaimed book *Mind and Matter,* which is included with the earlier treatise *What Is Life?* [5] in current editions. In his introductory comments he demonstrates a candor and courage that permeate his writing. "I can see no other escape from this dilemma [the dilemma posed by a single mind comprehending all existing data into a single whole] . . . than that some of us should venture to embark on a synthesis of facts and theories, albeit with second-hand and incomplete knowledge of some of them—and at the risk of making fools of ourselves.'' [6]

Schrödinger believed the Western world suffers from a massive, collective delusion, the assumption that the mind and consciousness are personal and individual:

> We have entirely taken to thinking of the personality of a human being . . . as located in the interior of its body. To learn that it cannot really be found there is so amazing that it meets with doubt and hesitation, we are very loath to admit it. We have got used to localizing the conscious personality inside a person's head—I should say an inch or two behind the midpoint of the eyes. . . . It is very difficult for us to take stock of the fact that *the localization of the personality, of the conscious mind, inside the body is only symbolic, just an aid for practical use*. (emphasis added) [7]

Where in the World Are We?

At first glance this view conflicts with the innate feeling that my ''I'' is behind my eyes, or at least somewhere in my head. This commonsense view of one's mind and conscious self as local, as occupying a particular place, gives rise naturally to the belief that we are observers looking outward from our bodies. This assumption has had an enormous force throughout our cultural history and lies at the heart of classical science, which contends that we can observe and measure from some outside vantage point, then think about what it all means. But in modern physics, this view, with Schrödinger's help, has been exploded.

Today most physicists believe that one simply cannot account for the findings of modern physics while holding to this view. The majority of them subscribe to the so-called Copenhagen interpretation of modern physics (so named because Niels Bohr, the primary architect of this view, was a Dane). According to this perspective, at the atomic level a real world simply does not exist until a measurement or observation is made. Prior to that time there is only a variety of possible outcomes for each subsequent event, each with its own probability of actually being realized once the observation is made. The observer—or a measuring device acting as his agent, according to some physicists—performs the pivotal act of "collapsing" all the coexistent possibilities into a single, coherent outcome that can only then be called an event. Prior to this time we are not justified in speaking of a real world of actual things and events, only of possibilities possessing a *potential* of being realized.

The eminent cosmologist and quantum theorist, John Archibald Wheeler, expresses the crucial role of the observer in the following way:

> An old legend describes a dialogue between Abraham and Jehovah. Jehovah chides Abraham, "You would not even exist if it were not for me!" "Yes, Lord, I know," Abraham replies, "but also You would not be known if it were not for me." In our time the participants in the dialogue have changed. They are the universe and man. [8]

Only by combining the observer and what is observed into a single whole does the current picture of the world make sense. This is one of the most radical differences separating the modern from the classical world view. The idea of an external, fixed reality pursuing its own course totally independent of an observer has been transcended in modern physics by one that incorporates mankind in an essential way. This essential feature of the new view is expressed by Professor Wheeler:

> Nothing is more important about the quantum principle than this, that it destroys the concept of the world as "sitting out there," with the observer safely separated from it. . . . To describe what has happened, one has to cross out that old word "observer," and put in its place the new word "participator." In some strange sense the universe is a participatory universe. [9]

It should be emphasized that the new views do not necessarily enshrine the sort of human consciousness that we value as a necessary component of this "observer" or "participator." As mentioned, some physicists maintain that a machine might do the observing just as well as a conscious being; others say that "pure, dumb awareness," perhaps not even of the human sort, would also qualify. Regardless of how this debate is resolved, what is important in our discussion is that current atomic science—the most accurate science mankind has ever devised—has gone beyond the notion of a fixed reality existing "out there."

This development—the coming together of the observer and what is observed—has affected both quantum mechanics and relativity, the two main branches of modern physics. The situation in relativity was expressed by scientist and philosopher Jacob Bronowski:

> Relativity derives essentially from the philosophical analysis which insists that there is not a fact and an observer, but a joining of the two in an observation . . . that event and observer are not separable. [10]

And Schrödinger expressed how this recognition applied to his field, quantum mechanics:

> Subject and object are only one. The barrier between them cannot be said to have broken down as a result of recent experience in the physical sciences, for this barrier does not exist. [11]

Schrödinger found in the ancient writings of India a great affirmation for this unitary point of view in which subject and object cannot be separated. He studied the Vedas, texts of wisdom that date in their earliest portions to 1500 B.C., and the Upanishads, a group of metaphysical treatises composed between 800 and 500 B.C. The primary question of these texts is, "What is the true nature of reality?" The answer given is that the Ultimate Reality one seeks is the same as the individual self. In the *Mundaka Upanishad:*

> Invisible, incomprehensible, without genealogy, colorless, without eye or ear, without hands or feet, unending, pervading all and omnipresent, that is the unchangeable one whom the wise regard as the source of all beings. [12]

Thus the individual self (Atman) and the ultimate reality (Brahman) one wishes to find constitute an indivisible whole. In the *Chandogya Upanishad:*

> That which is the finest essence—this whole world has that as its soul. That is Reality. That is *Atman*. That art thou [*tat tvam asi*]. [13]

In the Upanishads, Brahman is regarded as the source of everything. It is thus prior to time, space, matter, energy, and causality, and cannot be limited by them. And, if Atman, the individual self, and Brahman are one, this means that in some sense the self is the embodiment of perfection and cannot be affected by the ravages of space and time. Thus in the *Chandogya Upanishad:*

> The Self (*atman*) . . . is free from evil, free from old age, free from death, free from grief, free from hunger and thirst, whose desire is real, [and] whose thoughts are true. [14]

The implications of these views for immortality and for the oneness and unity of all individual selves deeply impressed Schrödinger, as we shall see, in large measure because they seemed so coherent with modern science. In this he was not alone. His contemporary, the famous physicist Niels Bohr, was similarly impressed with many implications of Oriental thought. Bohr was knighted by the Danish government and it thus became necessary for him to design a coat of arms. His choice for the central figure in the design was the sign of the Tao of ancient China, the embracing figures of yin and yang, which he felt symbolized one of his greatest ideas, the principle of complementarity.

Influenced by both physics and ancient teachings, Schrödinger came to believe that the mind could not be separated from the world and put in a box, the brain. Neither could the self be put in a body. Ultimately, the individual mind and self are not primary. In principle, they cannot be limited; they are intrinsically part of a larger whole.

One of the great perennial puzzles about the mind is this: Why, if there are many conscious egos, is there only one world that is concocted by all of them? Why not a different world for each person? Why do we not live in a Tower of Babel, each of us with a different world-picture, unable to communicate with each other?

There are at least two answers we can give, one of which was provided three hundred years ago by the philosopher Gottfried Wilhelm Leibniz with his doctrine of monads. Leibniz maintained that every monad is a single mind and is a world unto itself. It has no essential communication with any other monad. Agreement between them as to the nature of the world arises through some preestablished harmony. But this alternative, which satisfied Leibniz, will not do for Schrödinger, who calls it a "fearful doctrine" and "rather lunatic." One reason for his rejection of Leibniz's position is that it is inconsistent with modern science. But it also is not very satisfying for persons outside of science, even though it may be logically tidy. After all, most persons really *don't* feel they are incommunicado with everyone else; they *don't* feel walled off.

What other alternative is there to explain how a single vision of the world arises from minds that appear to be separate? Schrödinger gives this answer:

> There is obviously only one alternative, namely the unification of minds or consciousness. Their multiplicity is only apparent, in truth there is only one mind. [15]

Here Schrödinger takes us beyond the primacy of the person. Mind no longer is localized to the individual but is transpersonal, universal, collective—that is, it is *nonlocal*.

This problem was known long before Leibniz, of course. The Vedic philosophers of ancient India who penned the Upanishads grappled with it too. Of their view, Schrödinger says,

> The mystically experienced union with God regularly entails this attitude [the idea of the One Mind] unless it is opposed by strong existing prejudices; and this means that it is less easily accepted in the West than in the East. [16]

To emphasize this spiritual vision, he quotes the Islamic Persian mystic of the thirteenth century, Aziz Nasafi:

> On the death of any living creature this spirit returns to the spiritual world, the body to the bodily world. In this however only the bodies are subject to change. The spiritual world is one single spirit who stands like unto a light behind the bodily world and who, when any single creature comes into being, shines through it as through a window. According to the kind and size of the window less or more light

> enters the world. The light itself however remains unchanged. [17]

What evidence can be offered in favor of the One Mind? Its existence can be supported, says Schrödinger, by the inarguable fact that consciousness is *never* experienced in the plural, but *always* in the singular. ''Not only has none of us ever experienced more than one consciousness [at the same time], but there is also no trace of circumstantial evidence of this ever happening anywhere in the world.'' And this in spite of ''cases or situations where one would expect and nearly require this unimaginable thing to happen, if it can happen at all'' [18]—such as in cases of ''split personality,'' or multiple personality disorder as it is called by psychiatrists today.

As further support of a unified mind, Schrödinger cites the views on consciousness of Sir Charles Sherrington, one of the greatest figures in the history of neurological science. Sherrington charmingly described the brain as an ''enchanted loom'' and was led to a unitary view of the mind:

> Matter and energy seem granular in structure, and so does ''life,'' but not so mind. [19]

But in spite of his uncontested stature as a great scientist, Schrödinger does not believe that science alone can describe the ways in which minds actually become one. And neither does he believe Oriental mysticism alone holds the complete answer. Rather, *both* approaches will be needed. The answers will come, ''by assimilating into our Western build of science the Eastern doctrine of identity [of minds].'' [20]

The Immortal Mind

Schrödinger maintains that there are also valid reasons to believe that this One Mind is immortal. This assertion rests largely on new views of the nature of time. In modern physical science, an external time, like a completely objective world at large, does not exist. There is no evidence whatsoever that time is the entity portrayed in Newton's vision, nor has there ever been an experiment which shows that time flows. As we have noted, modern physics does away with the idea of the world-as-object, and along with it the view of time-as-object. There simply is no external, objective world in which an external, objective time could *be*. Thus, in modern physics not only are mind and world brought together in an essential way, so too are mind and time.

Now, if mind and time are dependent on each other, this raises the serious question of how time can possibly destroy the mind. Schrödinger's conjecture is that it cannot:

> I venture to call it [the mind] indestructible since it has a peculiar time-table, namely mind is always *now*. There is really no before and after for mind. There is only now that includes memories and expectations. [21]
>
> We may, or so I believe, assert that physical theory in its present stage strongly suggests the indestructibility of Mind by Time. [22]
>
> The fact remains that time no longer appears to us as a gigantic, world-dominating [force], nor as a primitive entity, but as something derived from phenomena themselves. It is a figment of my thinking. That as such it might some day put an end to my thinking, as some believe, is beyond my comprehension. Even the old myth makes Kronos devour only his own children, not his begetter. [23]

Schrödinger vividly describes a coevolution of mind and world, in which one is never around without the other—a possibility fully permitted if there is no before or after for the mind. During the process of evolution, he states:

> Only a small fraction of [species] embarked on "getting themselves a brain." And before that happened, should it all have been a performance to empty stalls? Nay, may we call a world that nobody contemplates even that? . . . a world existing for many millions of years without any mind being aware of it, contemplating it, is it anything at all? Has it existed? For do not let us forget: to say . . . that the becoming of the world is reflected in a conscious mind is but a cliché, a phrase, a metaphor that has become familiar to us. The world is given but once. Nothing is reflected. The original and the mirror-image are identical. The world extended in space and time is but our representation. . . . Experience does not give us the slightest clue of its being anything besides that. [24]

Our everyday mind, of course, is not adapted to thinking in the "now" mode that so impressed Schrödinger. That is one reason we prefer the linear, progressive, and chronological descriptions of classical science to the atemporal descriptions of modern physics. We are simply used to them. Consequently we make the most astonishing decision: We opt for a view of the world that is entirely objective and that contains an inexorable, one-way time—all this in spite of the fact that it entails death and destruction. We thus *exterminate* ourselves with our world view. Of this fact, we can only stand in sheer amazement.

Many eminent people in the West have come to views similar to Schrödinger's without resorting to physics or Eastern philosophy. Among them was Dostoevsky. In one passage in *The Possessed,* Stavrogin asks Kirilov whether he believes in a future eternal life, to which Kirilov replies,

> "No, not in a future eternal life, but in this present eternal life. There are moments—you can reach moments—when time suddenly stops and becomes eternal."

> "And do you hope to reach such a moment?"
>
> "I do."
>
> "It's hardly likely in our time," Stavrogin said slowly and thoughtfully, also without any irony. "In the Apocalypse, the angel promises that there'll be no more time."
>
> "I know. There's a lot of truth in it; it's clear and precise. When man attains bliss, there will be no more time because there will be no need for it. It's a very true thought."
>
> "Where will they hide time?"
>
> "Nowhere. Time is not a thing, it's an idea. It will vanish from the mind." [25]

In the Upanishads there is a corroborative analogy of Schrödinger's idea of immortality:

> The moon is the honey of all beings; all beings the honey of the moon. The bright eternal Self that is in the moon, the bright eternal Self that lives in the mind, are one and the same; that is immortality, that is Spirit, that is all. [26]

There are no doubt some intuitive or mystically minded readers who are sympathetic with Schrödinger's use of spiritual concepts to buttress his views of mind and immortality, but who cannot accept his use of modern physics: What can physics say about something as ethereal as consciousness, soul, and spirit? Physical science deals with the physical, and empirically minded persons are generally unwilling to believe that spirit and soul and mind are genuinely physical in nature, although they may manifest themselves in physical ways. If physics cannot deal with the nonphysical or the transphysical, why bother with it?

The reason is quite simple. The antipathy most persons feel toward physics is directed toward the physics of the past two centuries, not of this one. In our century something new has happened. As physicist Paul Davies puts it, as a consequence of the "bizarre and stunning new ideas about space and time, mind and matter, . . . [physicists] have learned to approach their subject in totally unexpected and novel ways that seemed to turn commonsense on its head and find closer accord with mysticism than materialism. [As a result,] science has actually advanced to the point where what were formerly religious questions can be seriously tackled." [27] But this does not mean that science in the end can possibly emasculate or control the spiritual. If we continually remember that the essence of mind and consciousness (as opposed to its physical manifestations) are safely beyond its reach, then we have nothing to fear by looking to its messages. And in some ways modern science has proved downright cordial to the eternal spiritual yearnings of mankind, as in its implications for unity, oneness, and nonduality of "man" and "world." As Schrödinger stated, the barrier between subject and object "does not exist." [28]

Schrödinger did, however, express reservations about the use of science to explore mind and spirit. He recognized that science, which goes far in illuminating reality, eventually reaches dark corners that it cannot brighten. Another source of light is then needed, a beam of understanding that cannot come from the methods which science employs. But from where will the alternative source of knowing come? Schrödinger answers:

> Our science—Greek science—is based on objectivation whereby it has cut itself off from an adequate understanding of the Subject of Cognizance, of the mind. But I do believe that this is precisely the point where our present way of thinking does need to be amended, perhaps by a bit of blood-transfusion from Eastern thought. That will not be easy, we must be aware of blunders—blood transfusions always need great precaution to prevent clotting. We do not wish to lose the logical precision that our scientific thought

> has reached, and that is unparalleled anywhere at any epoch. [29]

Schrödinger's assertion about the limitations of science in investigating consciousness has been rejected by many scientists, who describe a "cognitive revolution" in modern neurobiology, artificial intelligence, and psychology that has taken place since Schrödinger's time. Were Schrödinger alive today, his outlook might be different. These recent insights are proving it increasingly unnecessary, it is maintained, to employ concepts such as "spirit," "soul," or "mind." As a single example, in one of the most recent chronicles of the impressive advances in our understanding of human cognizance, Howard Gardner's admirable *The Mind's New Science: A History of the Cognitive Revolution,* the word "consciousness" is not even listed in the index. [30] And most of the references to "mind," which are listed, have to do with its similarities to a computer, the final reference being "*see also* brain." "Time" and "space" are entirely absent from the index, to say nothing of "immortality." Interestingly, in the discussion of the relevant historical ideas about cognizance, there is not a single reference to the philosophers of the East who, T. S. Eliot declared, make Western philosophers look like schoolboys.

Schrödinger believed that some questions about mind and spirit can never be answered scientifically, no matter how far-reaching the cognitive revolution. In principle, some matters are untouchable by empirical science. As he put it, ". . . we shall not expect the natural sciences to give us direct insight into the nature of the spirit." Our only hope, unless we wish to remain "cut . . . off from an adequate understanding of the Subject of Cognizance, the mind," is to find other paths to the knowledge we seek. [31]

The "All in All"

With the following sublime passage we come to a close of our look at Erwin Schrödinger's philosophy of mind. It is a fitting summary of his majestic, sweeping vision. Without prior knowledge it would be difficult to know whether the statement comes from a physicist, poet, or mystic. But, then, what *was* Schrödinger, if not all of them?

> A hundred years ago, perhaps, another man sat on this spot; like you, he gazed with awe and yearning in his heart at the dying light on the glaciers. Like you, he was begotten of man and born of woman. He felt pain and brief joy as you do. *Was* he someone else? Was it not you yourself? What is this Self of yours? . . . What clearly intelligible *scientific* meaning can this "someone else" really have? . . . Looking and thinking in [this] manner you may suddenly come to see, in a flash, . . . it is not possible that this unity of knowledge, feeling, and choice which you call *your own* should have sprung into being from nothingness at a given moment not so long ago; rather this knowledge, feeling, and choice are essentially eternal and unchangeable and numerically *one* in all men, nay in all sensitive beings. But not in *this* sense—that *you* are a part, a piece, of an eternal infinite being, an aspect or modification of it. . . . No, but inconceivable as it seems to ordinary reason, you—and all other conscious beings as such—are all in all. Hence this life of yours which you are living is not merely a piece of the entire existence, but is, in a certain sense, the *whole;* only this whole is not so constituted that it can be surveyed in one single glance. . . . Thus you can throw yourself flat on the ground, stretched out upon Mother Earth, with the certain conviction that you are one with her and she with you. You are as

> firmly established, as invulnerable, as she—indeed, a thousand times firmer and more invulnerable. As surely as she will engulf you tomorrow, so surely will she bring you forth anew to new striving and suffering. And not merely, "some day": now, today, every day she is bringing you forth, not *once,* but thousands upon thousands of times, just as every day she engulfs you a thousand times over. For eternally and always there is only *now,* one and the same now; the present is the only thing that has no end. [32]

Kurt Gödel

In the world of mathematics, Gödel is an almost mythical figure. In 1931, as a young Austrian mathematician, he stunned the world's community of logicians and mathematicians by providing two astonishing theorems, from whose effects the world of science is still reeling—the "incompleteness theorems," as they are called.

Gödel's first theorem is that any logical system that is complex enough to include, at least, simple arithmetic, can express true assertions that cannot be deduced from its axioms. And the second theorem says that the axioms in such a system, with or without additional supporting statements, cannot be shown in advance to be true from contradictions. What do these strange assertions mean? In short, they tell us two things: (1) a logical system that has any richness can never be complete, (2) nor can it even be guaranteed to be consistent. Gödel's proofs have been scrutinized by mathematicians and logicians of the highest caliber for over half a century and have not been shown to contain inconsistencies.

The meaning of Gödel's insights can be expressed in simple ways that are much less bewildering than his original proof. One example comes from the American humorist Ambrose Bierce, who expressed much the same conclusion in his definition of the mind that he included in *The Devil's Dictionary*. The mind, he said, is a substance secreted deep within the brain, whose chief preoccupation is thinking about itself—the problem being that it has only itself to think about itself *with*. It would be difficult to illustrate the problem posed by Gödel more succinctly. In order to think about the mind one has to employ the mind; one has to step outside the mind. In this situation the mind must function simultaneously as subject and object. But a dilemma arises: the mind is then incomplete because something has been withdrawn, which will affect any future observation about it. Yet if one does not step outside the mind to observe it, the observation is not possible. By the inherent process of thinking about the mind *with* the mind, both the completeness and the consistency of our reasoning are bound to suffer.

These observations strike at the heart of science's ideal aim, which is to devise a complete and consistent picture of nature. Gödel showed that this cannot be done. Not only is it not technically possible to accumulate all the data necessary to formulate such a picture, the very goal itself is hopeless. Gödel's theorems show that nature's laws, if they *are* consistent, as we believe them to be, must be of some inner formulation quite different from anything that we now know, as Jacob Bronowski put it, and which at present "we have no idea how to conceive." [33]

Gödel's proofs are roundly neglected by practical-minded, journeyman scientists, but they have commanded the attention of all major thinkers who are concerned with the conceptual foundations of science. As mathematician Rudy Rucker puts it in his book, *Infinity and the Mind,* "Kurt Gödel was unquestionably the greatest logician of the century." [34] In Rucker's view he also may have been one of our greatest philosophers. When Gödel died in 1978, one of the speakers at a memorial service compared him not only with Einstein, but with the Austrian writer Franz Kafka—because of the almost sinister and forboding incompleteness theorems we have briefly described above. They are indeed Kafkaesque—rational thought telling us that rational thought can never penetrate to the final, ultimate truth.

If this seems terribly depressing, it need not be. Rucker expresses the surprising, freeing experience that can actually come from immersing oneself in Gödel's ideas and reasoning one's way to his conclusions:

> Paradoxically to understand Gödel's proof is to find a sort of liberation. For many logic students, the final breakthrough to full understanding of the Incompleteness Theorem is practically a conversion experience. This is partly a by-product of the mystique Gödel's name carries. But, more profoundly, to understand the essentially labyrinthine nature of *the castle* is, somehow, to be free of it. [35]

Rucker visited Gödel on a series of occasions beginning in 1972, and has given us an intimate picture of him. He was a small man. His voice had a high, singsong quality. "He frequently raised his voice toward the ends of his sentences, giving his utterances a quality of questioning incredulity. Often he would let his voice trail off into an amused hum. And, above all, there were his bursts of complexly rhythmic laughter. The conversation and laughter of Gödel were almost hypnotic. Listening to him, I would be filled with the feeling of perfect understanding." [36]

Although he was not Jewish, Gödel was forced to flee Europe because of the Second World War. He came to Princeton to the Institute of Advanced Studies. There he met Einstein and the two of them spent much time together. But Gödel spent most of his life in solitude. He worried a good deal about his health and is known to have always kept himself warmly bundled up. He became increasingly reclusive, even secretive in his later years.

It is easy to think that such a person would be extremely unlikely to embrace the idea of a universal, nonlocal mind. But in Rucker's last encounter with Gödel, by telephone in 1977, some ten months before his death, Gödel conveyed his soaring ideas about the mind:

> I [Rucker] had been studying the problem of whether machines can think. . . . In short, I had begun to think that consciousness is really nothing more than simple existence. By way of leading up to this, I asked Gödel if he believed there is a single Mind behind all the various appearances and activities of the world.
>
> He replied that, yes, the Mind is the thing that is structured, but that the Mind exists independently of its individual properties.
>
> I then asked if he believed that the Mind is everywhere, as opposed to being localized in the brains of people.
>
> Gödel replied, "Of course. This is the basic mystic teaching."
>
> . . . "What causes the illusion of the passage of time?"
>
> Gödel spoke not directly to this question, but to the question of what my question meant—that is, why anyone would even believe that there is a perceived passage of time at all.
>
> He went on to relate the getting rid of belief in the passage of time to the struggle to experience the One Mind of mysticism. Finally he said this: "The illusion of the passage of time arises from the confusing of the *given* with the *real*. Passage of time arises because we think of occupying different realities. In fact, we occupy only different givens. There is only one reality." [37]

Albert Einstein

When Einstein published his special theory of relativity in 1905, the world was never to be the same. Einstein shattered the fundamental, self-assured foundations of the classical physics of Newton—its deterministic, causal framework; its linear, flowing time and empty space; its rigid compartments of matter and energy. But these insights are not our concern here. We want to look at some of his unconventional views about man's place in the world, and how he saw the mind.

A Living Physics

One of the most remarkable facts about Einstein was the way in which he seemed to *live* his physics. His example totally demolishes the idea that physics is purely an intellectual affair. In the early days when he was attempting to reason through what later became the special theory of relativity, we get the picture of a man almost tormented with the force of his ideas. Dealing with them was a great physical strain. In fact, this mental overload nearly caused physical illness and he frequently had to retreat to the country for a respite in order to regain his strength. This is not the picture of cold intellection; this is a man being moved by powerful living forces.

No one can read accounts of his life and not be struck by the sense of holiness with which he regarded all of creation. Physics was for him no dull affair; it was an attempt to understand God's work. Einstein's theology and his science were tied hand-in-glove; that they could exist separately seemed unthinkable to him.

Telling God What to Do

He shares this attitude with his predecessor, Sir Isaac Newton. Both these men stand in stark and paradoxical contrast to the ideal that has come to prevail in modern science: that it should be purged of all tinges of God and divinity. Nature is dispassionate, we are now told; it simply is what it is. The "divine stamp" or the "clockmaker" who fashioned the clock are our own foolish projections, it is said. But Einstein knew better. His goal in understanding any great problem was always to ask himself: How would God think? This led on one occasion to a remonstration by a colleague that he should stop telling God what to do!

Science has strange ways in dealing with persons such as Einstein. It picks and chooses among their insights. Those that are fashionable or permissible are selected for scrutiny, and those that are not are rejected. This is always defended with the reason that many of the ideas, even of great men, are scientific and valid while others are not; we are not obliged to take the whole package. Certainly caution is necessary, and science must proceed by trusting only proof, not authority. But the issue is more complex than that, because the reasons that are sometimes given for rejecting "part of the package" are entirely unscientific, and commit the same fallacies that they claim to oppose.

Take the case of Newton. He believed in God with a fervor. Newton spent much of his energy investigating alchemy and he was convinced that his reputation would eventually rest more on his alchemical contributions than on his physics. Yet all of Newton's leanings toward God and alchemy have been forgotten, and physics as a whole has been purged of such things. The science that followed Newton came to regard nature as a neutral, godless machine. Machines are deterministic and run according to the ironclad, unswerving laws of nature. In this view there is no room for the heavily anthropomorphic, God-permeated alchemy of Newton, thus these views were gradually filtered from his legacy. The remaining residue had to be compatible with the neutral, value-free, machinelike picture of the world that completely dominated science thereafter.

But the paradox here, expressed by the biologist Rupert Sheldrake, is this:

> What could be more anthropomorphic in human modelling than to say that everything's a machine? Machines are entirely and specifically human creations, . . . based on a particular kind of human activity. [38]

And as physicist David Bohm has put it,

> [The model of the machine] is based on projecting into matter the sort of things that you do with your body. [39]

The gods never vanish from our images of nature. Our nature is still godlike, only the god is not Newton's but our own current object of adoration—the machine (in the case of the universe) or the computer (in the case of the brain).

How did the greatest scientist since Newton see himself in relation to other persons? The answer is not simple. If we are to judge from his social behavior, Einstein seemed to possess a strong sense of inner isolation. As he described this feeling,

> Arrows of hate have been shot at me too, but they never hit me, because somehow they belonged to another world with which I have no connection whatsoever. I live in that solitude which is painful in youth, but delicious in the years of maturity. [40]

And,

> My passionate sense of social justice and social responsibility has always contrasted oddly with my pronounced lack of need for direct contact with other human beings and human communities. I am truly a "lone traveler" and have never belonged to my country, my friends, or even my immediate family with my whole heart; in the face of all these ties, I have never lost a sense of distance and a need for solitude—feelings which increase with the years. [41]

Even when he was among others, Einstein could still be alone with his thoughts. His intimates, such as the physicist Leopold Infeld, spoke of the "unbelievable obstinacy" of his thought processes, which he could exercise even in public. His son-in-law provides this observation about this capacity:

> Regardless of how many surround him, he is always alone—not lonely, but alone. Einstein does not need people. He receives them with warmth and kindliness, but they are in no way necessary to him. You see this in his eyes when he leaves them; the expression is already one of extreme contemplation, and he is barely aware he has been with them; the line of his thinking is unbroken. [42]

Not only did Einstein set himself apart from the world socially, he distanced himself from the world in another way. For him, the universe is only partly knowable, and this fact alone sets us apart from it in a certain sense. As he expressed this feeling,

> Out yonder there was this huge world, which exists independently of us human beings and which stands before us like a great, eternal riddle, at least partially accessible to our inspection and thinking. [43]

As he makes clear, Einstein believed in a reality independent of human knowledge. But although it is distant from us, it is intrinsically harmonious and lawful. This makes it knowable to some extent by the "free constructions" fashioned by the human mind, which begin and end in experience. The fact that the world *is* knowable, if only partially, was a never ending source of astonishment and awe for Einstein.

On the surface there seems little connection here with the philosophy of Schrödinger or Gödel. They seem to be describing different worlds. Einstein's world does not seem the kind of place where the sort of nonlocal mind and being we have focused on so far could take root and flourish. In fact, it seems decidedly local; it is a world apart from us, a world that accentuates our aloneness and isolation, in the everyday world of objects.

But that is not the whole story. This very private man, who firmly believed in a reality independent from us, demonstrates at other moments elements in his life and thought in which an unmistakable oneness shines through. For instance, when asked during a serious illness whether he was at all afraid of death, Einstein replied,

> I feel such a sense of solidarity with all living things that it does not matter to me where the individual begins and ends. [44]

Where *did* the individual begin and end for Einstein? The boundaries of the person were seemingly far-flung. We get a hint of this view in his attitude about freedom of the will, in which he reveals his belief that we have unseverable ties with all the things and events of the world—an affinity which is so intimate that the entire question of individual freedom is nonsensical. Our concept of freedom of the will in one sense is very limited, implying an isolated individual situated in the here-and-now who can exercise it. Einstein does not share this local concept. For him, freedom of the will is tied to an endless chain of events extending far into the past in an indefinitely large expansion. He expressed it this way:

> Honestly I cannot understand what people mean when they talk about the freedom of the human will. I have a feeling, for instance, that I will something or other; but what relation this has with freedom I cannot understand at all. I feel that I will to light my pipe and I do it; but how can I connect this up with the idea of freedom? What is behind the act of *willing* to light the pipe? Another act of willing? Schopenhauer once

> said, "Man can do what he wills but he cannot will what he wills." . . . when you mention people who speak of such a thing as free will in nature it is difficult for me to find a suitable reply. The idea is of course preposterous. [45]

What Einstein seems to renounce is the tenacity with which people cling to the idea of an isolated, primary ego, an "I" around which everything revolves. But the ego, which craves its freedom to will, is not the measure of how the world works. Willing is *not* entirely focused in the here-and-now, and it is *not* concentrated and culminated in me at this moment. In this view Einstein does not forfeit his ability to act: he can *still* light his pipe. What he does renounce, I believe, is the pathological, morbid, *local* focus on the ego as the definitive atom of humanity.

Einstein furthermore takes a massive step beyond the isolated individual as the sole arbiter of the world picture in his theory of special relativity. In it the world picture cannot be understood by supposing that there is a set of events that can be uniformly observed by all independent observers. Rather, the world must be understood not as a collection of objective, external events that are the same for everyone, but as a set of *relationships*. Indeed, "event" and "individual observer" cannot be defined independently of each other in the new view. Through his insights a new unit in physics came into being, as Bronowski puts it, which is made up of an inextricable triad: the event, the observer, and the signal connecting them:

> Physics does not consist of events; it consists of observations, and between the event and us who observe it there must pass a signal—a ray of light perhaps, a wave or an impulse—which simply cannot be taken out of the observation. . . . Event, signal and observer: that is the relationship which Einstein saw as the fundamental unit in physics. Relativity is the understanding of the world not as events but as relations. [46]

Thus in special relativity individuals participate in the world picture only by being included in something larger than the individual self. Unless one goes beyond the self there *is no* world picture. This stunning theory, which has been proved beyond all doubt and which is universally accepted, is quite harmonious with Einstein's belief that individuals fit into ever-larger units—how, as he put it, they are connected "with all living things."

In his personal religious views Einstein takes definite steps away from the preeminence of the person toward the Whole, the One. This is yet another expression of transindividuality and the nonlocality of person in his thought. As he commented,

> The individual feels the futility of human desires and aims and the sublimity and marvelous order which reveal themselves both in nature and in the world of thought. Individual existence impresses him as a sort of prison and he wants to experience the universe as a single significant whole. [47]

Going beyond the prison of individuality to an experiential awareness of this "significant whole" was for Einstein a great life-task, which he describes in an oft-quoted remark:

> A human being is part of the whole, called by us "universe," a part limited in time and space. He experiences his thoughts and feelings as something separate from the rest—a kind of optical delusion of his consciousness. This delusion is a kind of prison for us, restricting us to our personal decisions and to affection for a few persons nearest us. Our task must be to free ourselves from this prison by widening our circle of compassion to embrace all living creatures and the whole of nature in its beauty. [48]

Einstein at certain moments emphatically affirms the value of breaking the bondage of the personal ego, the sense of the local "I." In doing so, his philosophy sounds decidedly Eastern:

> The true value of a human being is determined primarily by the measure and the sense in which he has attained liberation from the self. [49]

Yet Einstein always comes back to Earth after these mystical flights, and his vision is invariably and refreshingly supplemented by the realization of the importance of the here-and-now. For instance, it is never enough to dwell only in the spiritual domain, for

> . . . those who would preserve the spirit must also look after the body to which it is attached. [50]

In fact, the similarity between the Eastern and the Einsteinian view of the relationship of the liberated self (which is part of the Whole) and the local, limited person, is striking. Compare his comments above with those of Lama Govinda, a modern Buddhist scholar:

> [Our essential oneness with the universe] is not sameness or unqualified identity, but an organic relationship, in which differentiation and uniqueness of function are as important as that ultimate or basic unity. . . . Individuality and universality are not mutually exclusive values, but two sides of the same reality, compensating, fulfilling, and complementing each other, and becoming one in the experience of enlightenment. This experience does not dissolve the mind into an amorphous All, but rather brings the realization that the individual itself contains the totality focalized in its very core. [51]

There is an unmistakable intimation of immortality in Einstein's views, views that flow from the interpretations of time in modern physics, in which it is deprived of its external, flowing, linear character. Based on these views, entirely new meanings come about for past, present, and future; and the concept of death as an ultimate, final event breaks down.

In 1905 Einstein published his treatise on special relativity. At the end of this paper, which was to change forever our concepts of space and time, he thanked his close friend Michele Besso. It was Besso, when they were workers together in the patent office in Bern, with whom he had agonized over these seminal ideas. Their profound friendship endured over a lifetime, and in 1955 when Besso died, Einstein wrote a letter to Besso's surviving son and sister that expresses his view on immortality:

> The foundation of our friendship was laid in our student years in Zurich, where we met regularly at musical gatherings . . . later the Patent Office brought us together. The conversations during our mutual way home were of unforgettable charm. . . . And now he has preceded me briefly in bidding farewell to this strange world. This signifies nothing. For us believing physicists the distinction between past, present, and future is only an illusion, even if a stubborn one. [52]

We do not find in Einstein's thought the mystical flights of Schrödinger, yet his philosophy still soars to great heights. The "cosmic religious feeling" of which he frequently spoke is exactly that—*cosmic,* in the literal sense of the word. In spite of his seeming aloneness and need for solitude, in his philosophy he does not deify the individual. There is more than the person: There is the Whole, of which the individual is a part. And there is immortality.

These scientists—Schrödinger, Gödel, Einstein—tug us toward the same territory, if with different tenacity. They tell us that there is more than the individual person and the here-and-now. In their own way, they each tell us that our commonsense interpretations of our mind and self are seriously flawed. Reality—including time, space, person, mind, life, and death—is not what we have always taken it to be.

6

Mind and Quantum Physics

The universe shows evidence of . . . mind on three levels. The first level is the level of elementary physical processes in quantum mechanics. Matter in quantum mechanics is not an inert substance but an active agent, constantly making choices between alternative possibilities according to probabilistic laws. Every quantum experiment forces nature to make choices. It appears that mind, as manifested by the capacity to make choices, is to some extent inherent in every electron. The second level at which we detect . . . mind is the level of direct human experience. Our brains appear to be devices for the amplification of the mental component of the quantum choices made by molecules inside our heads. . . . There is evidence . . . that the universe as a whole is hospitable to the growth of mind. . . . Therefore it is reasonable to believe in the existence of a third level of mind, a mental component of the universe. If we believe in this mental compo-

nent of the universe, then we can say that we are small pieces of God's mental apparatus.

—FREEMAN DYSON [1]

Maybe a fella ain't got a soul of his own, but on'y a piece of a big soul—the one big soul that belongs to ever'body.

—JOHN STEINBECK
The Grapes of Wrath

Henry Margenau and the Universal Mind

It is easy to find support for the soul-like, nonlocal nature of the mind among poets, mystics, and philosophers, and one can even add to the list a few scientists who have toyed with the idea from time to time. But it is exceedingly rare to find a contemporary scientist, distinguished among his colleagues for making fundamental contributions to his discipline, who has openly declared that the mind is universal. Such a person is Henry Margenau, Professor Emeritus of Physics and Natural Philosophy at Yale University. Following a career as a distinguished theoretician in both molecular and nuclear physics, Professor Margenau began an investigation of the philosophical foundations of natural science.

Today most physicists lead somewhat curious lives. They regard physics in an almost completely utilitarian way—as a tool, as a means to an end. This way of looking at physics leads to a definite duplicity of existence. The world view of physics is radically different from that in which the physicist—and the rest of us—lives outside the laboratory. When the physicist lays down his tools at the end of the day, he leaves behind the world view required to use these tools in the laboratory, and dons another dominated by common sense.

Of course, all persons have multiple world views that guide them in different circumstances, as the example of the businessman who may be godfearing on Sunday but ruthless and uncompassionate in his professional dealings on other days. And in any case, multiple views of reality are absolutely necessary to remain alive and function in this world, as psychologist Lawrence LeShan has convincingly argued in his book, *Alternate Realities*. [2] The task is always to choose the way of looking at the world that best fits the situation. It would be an impossible feat, and extremely unwise, to live consistently according to a single world view. In our blood we know this, and none of us tries to hang on to the same view of reality all the time. When we work, play, and dream we are constantly moving in and out of different assumptions about reality, although we may not do so consciously.

Professor Margenau has expanded the world view of the physics laboratory by implying that physics hints at a reality meaningful not only when one is doing physics but when one is bidding on the floor of the stock exchange, crossing the street, or cleaning the refrigerator. This world view is far-reaching and all-inclusive, containing all the subsidiary, utilitarian world views we move in and out of in our daily lives.

Many books have appeared recently dealing with the "new physics" and how this body of knowledge may relate to human concerns. But Margenau's contribution is unique among them. His book, *The Miracle of Existence,* [3] is the most robust and breathtaking statement to appear in many years, perhaps since Schrödinger's proposals. The spiritual overtones come through unmistakably, for this is not just a book about physics, it is a book about God.

Margenau uses the term "Universal Mind," which is equivalent to our theme of nonlocal mind. We have chosen the latter throughout this book because it may contain a neutrality that "Universal Mind" may lack, the latter having become freighted with many religious connotations through the years. But in the following discussion we will primarily follow Margenau's own terminology. The reader, however, may safely interchange ''universal'' with ''nonlocal'' and do no injustice to Margenau's lofty views.

''Unity'' and ''oneness'' are a staple in the everyday diet of the poet and the mystic. But why can physicists speak today about such matters? Have our physicists ''gone mystic''?

One of the reasons physicists are justified in addressing these issues is a ''fact so peculiar yet commonplace that it fails to amaze the modern scientist[:] . . . the sameness of the properties of the elementary constituents of matter.'' [4] Every grade school pupil knows, says Margenau, that there is an amazing consistency and unity in nature. All oxygen atoms, and all atoms of a given species, have the same mass or weight. All electrons have the same mass, spin, and charge, which reflects an accuracy that could never be achieved in man-made things, and this is true for all the properties of the known elementary constituents of matter. When we find this sameness and unity in the macroscopic, everyday world—for example, the fact that coins and paper bills are of the same value, that automobiles are of the same make, or that all machinery from an assembly line is uniform—we at once assume that they were designed by man and therefore reflect the underlying mind of man. ''Should we not make a similar assumption, and are we not compelled to make it, with respect to the fundamental entities of atomic and nuclear physics . . . [though] the intelligence behind them was not that of man[?]'' [5]

Margenau acknowledges the obvious resistance, emotional as well as rational, that the Western scientist feels in postulating a Universal Mind of which every conscious being and ''perhaps every entity composing the world'' is a part. [6] Wishing to make this idea ''less repulsive to the modern physicist,'' he therefore makes a number of further observations that he feels may soften the blow.

There is a feature of recent nuclear physics contained in current gauge theory that strongly hints at a kind of oneness in the physical world. In explaining it Margenau makes use of the term *onta,* which he uses to designate "any entity whatsoever, especially when it defies ordinary intuition." [7] (*Onta* is the plura of *on,* from the Greek word for "being.") The findings from nuclear physics suggest that under certain situations different *onta* have a way of coming together and losing their identity—a definite suggestion that oneness does exist at the most basic levels of nature. But this form of oneness is complex; for although there is a loss of identity, the *onta* do not merge into total indistinguishability; they still retain their "number." Thus individuality persists, paradoxically, within the movement to unity at this level of nature.

As an example, consider the behavior of neutrons and protons. When they are separated in space and are therefore not interacting, one is neutral and the other carries a positive charge. But when they come sufficiently close together "their identities disappear, their properties merge, and a distinction between them becomes impossible. But they are still two *onta.*" [8]

Furthermore, Margenau points out that scientists today assume three different fundamental forces among the constituents of the nuclear world. The strength of these forces depends on the energy of the interacting entities. At small energies of interaction the strength of these forces differs enormously. "Strangely, however, the three forces approach equal values at extremely high energies if current theories are correct." [9] Again, a hint of oneness.

What are we to make of these observations? Atoms are not humans, and electrons are not minds. Can we even compare the "oneness" that may exist at the physical and the mental levels? Margenau is certain that one *can* speak as a physicist about oneness that encompasses distant levels of nature because of the revelations of modern science. As evidence that he is not alone in this belief, he enlists the insights of two central figures in modern physics, Werner Heisenberg and David Bohm. Shortly before his death Heisenberg published a paper which contained the proposal that certain fundamental, mechanistic, commonsense concepts such as "being composed of" and "having distinct and namable parts" may be meaningless for the ultimates with which physics seeks to deal. And physicist Bohm expressed the same sentiment. "Thus," he said, "one is led to a new notion of *unbroken wholeness* which denies the classical idea of analyzability of the world into separately and independently existent parts." [10] And if "parts thinking" has broken down at the level of atoms, Margenau asks a central question:

> Should this kind of denial [the denial of separability into parts] also be necessary for consciousness, for mind, so that the question of separate minds making up or adding to the universal mind could become meaningful? Some philosophers who contributed to the Vedas and the Upanishads would clearly answer in the affirmative, and the claims of mystics to have merged with God in ecstasy provide further evidence for the numberless nature of souls. [11]

These great physicists are implying that the concept of wholeness is not limited to the atomic domain. If "parts thinking" is inappropriate at the level of atoms, it is also malapropos at the level of minds. And what is a mind without parts? It is the One Mind or the Universal Mind, the "Tao, Logos, Brahman, Atman, the Absolute, Mana, Holy Ghost, Weltgeist, or simply God."

For Margenau, the fact that we perceive the same world is evidence for the existence of the Universal Mind. Granted, everyone's vision of things is not precisely identical, a fact that is amply documented by decades of experiments in perceptual psychology. Yet there is a rough equivalence between our visions that no one can doubt; we can communicate shared experiences about our world without too much difficulty. Now, what are we to make of the fact that we collectively share a coherent picture of the world? This fact is profoundly important, says

Margenau. After we take in incoming stimuli, they are finally, he says,

> transcribed . . . [into a] physical reality, in essence the same for all. . . . [This] oneness of the all implies the universality of mind if we remember that matter is a construct of the mind. [12]

This significant possibility is overlooked consistently by perceptual psychologists, neurologists, and philosophers of the mind. If, as modern neuroscience agrees, we know nothing except through the senses, then why is there not a different world for each brain? Brains are *not* alike even in identical twins. And the same brain, from one moment to the next, can perceive the same stimuli in a different way, and make a different world picture. When we consider how radically different the pictures that our brains make *could be,* it is astonishing that our world pictures turn out to be as coherent as they are.

And the reason they *are* coherent, Margenau implies, is *not* because our brains are similar or work the same, *but because our minds are one*. It takes a single consciousness to make a single picture of the world, especially when that world picture is being assembled by some 5 billion brains on this planet. Only the One Mind, a Universal Mind, could do such a thing. To perform in such a way it must be nonlocal in the sense of being beyond individual brains and bodies. If the One Mind were not at work shaping the vast amount of sensory data processed every moment by the sea of brains on the Earth, we might expect world pictures to be formed that are so disparate as to be incommunicable.

Some counter that the pictures we make of the world are one because there is only one world to make the picture *from*. This view is that of naive realism, and Margenau and modern physics in general ask us to go beyond it, for there is really no "out there" that we can regard as totally external, objective, and the same for everyone. There is an aspect of reality that is deeper than the "outside" objects, and must include the mind. Ultimately this is the reality of the One, the Universal Mind, which in its most comprehensive expression is God.

Even if we recognize our participation in Margenau's Universal Mind, we are still habitually tempted to set ourselves apart from it. We persistently tend to think about the One in the same way we think about objects—a tree, a rock, or the sunset. The One is likely to become a thing for us just like any other. And so we seize the concept of the Universal Mind and reify it as the One Great Thing. But the One cannot be the equivalent of the One Thing. That is "parts thinking." In the ancient messages of the Vedas and the Upanishads, to which Margenau as well as Schrödinger refers, the One is never used as if it implies The Biggest Thing in the World. It is used metaphorically. Ken Wilber's view is helpful and clear on this important point:

> Speaking of Reality as the One . . . is undoubtedly helpful, for it *metaphorically* points to reality as that "single" and absolute ground of all phenomena—[but] it is helpful *provided* we remember that it is metaphor. . . . So wherever . . . [spiritual] traditions speak of the "one," they . . . mean not literally "one," but what could better be expressed as "non-dual." This is not a theory of monism or pantheism, but an experience of nonduality. [13]

The vision of the One Mind is consistent with modern physics and with many of humanity's great spiritual traditions. And if the One is truly that, if *it is* One, then we are part of it—but not only a "part," for then we violate the nondual relationship that Wilber warns against. Ultimately we must go beyond the idea that our mind is a part of anything else—recognizing, as the physicist Schrödinger said, that at some level we *are* the One Mind. For if anything were outside of it, including ourselves, it could not be the One—total, complete, ultimate. Margenau, like Schrödinger, is clearly aware of the spiritual implications of the absorption of the "part" in the whole. Nothing less than mankind's relation to God is concerned. As he says,

> If my conclusions are correct, each individual is part of God or part of the Universal Mind. I use the phrase "part of" with hesitation, recalling its looseness and inapplicability even in recent physics. Perhaps a better way to put the matter is to say that each of us is the Universal Mind but inflicted with limitations that obscure all but a tiny fraction of its aspects and properties. [14]

The Mind as a Field

In life sciences such as biology and medicine, scientists are not used to dealing with nonmaterial entities. The very phrase conjures thoughts of ghosts and spirits. But in modern physics the situation is different. Here scientists have concepts for many nonmaterial entities, many of which are called *fields*. While these are not material, most of them are nonetheless associated with matter. These include, for example, the flow field of a moving fluid, the electric and magnetic fields surrounding bodies, the temperature field of the atmosphere, and the stress field within a compressed solid.

But there are some fields that do *not* require the presence of matter to be meaningful. They are not attached to material things and could never be called material—for example, the metric field in general relativity, radiation fields, and several abstract fields occurring in nuclear physics. In addition there are the probability fields, which are among the basic "observables" in quantum physics, says Margenau, along with quantities like position, velocity, mass, and energy. Probability fields characterize the essence of quantum physics, and play a key role in Margenau's model of the Universal Mind.

The idea that the mind could be a nonmaterial field capable

of causing physical changes to occur in the world has not been warmly received in modern biology. Biologists are only reluctantly beginning to realize, if at all, that the mind is not physically dependent on the brain and body, and that it will not be understood completely in terms of the brain's chemistry and anatomy. British physicist Paul Davies observes, ''Physics, which has led the way for all other sciences, is now moving towards a more accommodating view of mind, while the life sciences, following the path of last century's physics, are trying to abolish mind altogether.'' [15] Davies cites the observation of bioscientist Harold Morowitz about this ''curious reversal'':

> What has happened is that biologists, who once postulated a privileged role for the human mind in nature's hierarchy, have been moving relentlessly toward the hard-core materialism that characterized nineteenth-century physics. At the same time, physicists, faced with compelling experimental evidence, have been moving away from strictly mechanical models of the universe to a view that sees the mind as playing an integral role in all physical events. It is as if the two disciplines were on fast-moving trains, going in opposite directions and not noticing what is happening across the tracks. [16]

Although heretical when viewed against the backdrop of modern biological materialism, Margenau's views on the mind's nonmaterial nature would find sympathy with some of this century's greatest physicists. Niels Bohr stated, ''We can admittedly find nothing in physics or chemistry that has even a remote bearing on consciousness.'' [17] And his contemporary, Werner Heisenberg, the originator of the Uncertainty Principle in modern physics, also put the matter bluntly. ''There can be no doubt,'' he said, ''that 'consciousness' does not occur in physics and chemistry, and I cannot see how it could possibly result from quantum mechanics.'' [18] In their own way, these physicists—Bohr, Heisenberg, and Margenau—occupy essentially the same position: Consciousness cannot be fully accounted for by the physical sciences as they are currently understood.

Many biologists have gone beyond the radical version of materialism enunciated in 1750 by the influential French scientist Julien de La Mettrie. "Let us conclude boldly then," he confidently urged, "that man is a machine." [19] Currently one of the favored ways of transcending this extreme position is through the philosophy of emergent materialism, according to which the mind "emerges" at a certain level of complexity of brain function. But this is still a materialistic way of thinking, for the mind remains dependent on the brain for its origin, as a baby on its mother for its birth. "Emergence" essentially allows biology to slip back into materialism through the back door.

Even if the mind does indeed "emerge" in some sense from the brain, one may wonder what is ultimately explained by the word "emerge." The problems surrounding the philosophy of emergence are clearly posed by the philosopher of science, Sir Karl Popper. "From an evolutionary point of view," he states, "I regard the self-conscious mind as an emergent product of the brain. . . . [But] I want to emphasize how little is said by saying that the mind is an emergent product of the brain. It has practically no explanatory value, and it hardly amounts to more than putting a question mark in a certain place in human evolution." [20]

A major reservation many biologists and philosophers have about Margenau's views concern the nonmaterial nature he attributes to the mind. How can a nonmaterial entity that is totally independent of matter *do* anything? Can nonmaterial "things" act on material ones? Margenau is suggesting the irrational possibility that ghostly things can push material things around. But in quantum mechanics, the irrational has indeed come to pass: interactions between the nonmaterial and the material are commonplace. As Margenau states, "interactions between [the] immaterial and [the] material are known to occur, indeed abound [in modern physics]. Every electric motor depends on it. . . .

[And] elusive entities such as probability fields, a purely mathematical construct, . . . affect the behavior of atomic entities. [21]

How might this two-way interaction between nonmaterial mind and material brain take place? In the typical situation, when work is done between two interacting entities some amount of energy is transferred between them. But not all interactions occurring in the physical world are of the nature of energy exchanges. There are control or guidance mechanisms, for example, in which no work is done and hence no energy is transferred. Another example Margenau cites is a train going around a curve: The rails press against the wheels of the car, exerting a force, but the force is at right angles to the displacement, and thus no work is done. Here it is fundamental to realize that ordinary, common notions of what it means for one system to "do work" on another can break down.

When entities interact in nature, the total energy of the interacting system before and after the interaction takes place remains the same (energy is "conserved"). But in the world of quantum physics, Margenau states, this does not always hold true:

> There are instances in which the principle of conservation of energy in its customary form does not hold: as examples one could cite the passage of electrons through barriers. . . . and perhaps most miraculous of all, is that a physical mass can be created out of nothing without contradicting the laws of physics. [22]

This opens a wide door for the action of the nonmaterial mind on the material brain, without the mind having to come up with some quotient of energy to expend in the process. The picture that emerges from Margenau's observations, then, is this: *The nonmaterial mind may be completely free and independent from the physical brain, yet fully capable of influencing it, without having to furnish any of the energy required in the interaction between the two.* This possibility, long denied in biology, is fully permissible in modern physics, Margenau contends, in perfect accordance with known principles and without violating any laws. He enlarges on how the mind-brain interaction might come about:

> In very complicated physical systems such as the brain, the neurons, and sense organs, whose constituents are small enough to be governed by probabilistic quantum laws, the physical organ is always poised for a multitude of possible changes, each with a definite probability; if one change takes place that requires energy . . . the intricate organism furnishes it automatically. Hence, even if the mind has anything to do with the change, that is, if there is mind-body interaction, the mind would not be called on to furnish energy. [23]

Thus the answer to the perennial puzzle of how the nonmaterial mind furnishes energy to affect the material brain or body may be: It doesn't. The energy can come from the brain.

Through the Glass Darkly: The Time Slit and the Personal Wall

What would a Universal Mind "look" like? Of this Mind, Margenau states,

> Its knowledge comprises not only the entire present but all past events as well. Much as our thought can survey and come to know all space, the Universal Mind can travel back and forth in time at will. [25]

If our minds are part of this Universal Mind, then they, like it, are also nonlocal in time and space. But if this is so, why do we feel so *local*? Why do we have such an overpowering sense of the present and such a sense of confinement to this immediate space? Why do we feel so *individual,* locked inside our body? Why should we be so tormented, as we have been for millennia, by the question of whether we really have any freedom of choice, or whether the game is fixed from the beginning? These are not characteristics we would expect of minds that are a part of a nonlocal, Universal Mind.

Margenau believes our sense of our own universality is obscured by the physical constraints of the body. Yet these physical limitations are not absolute, and there are many people throughout history who have managed to go beyond them. All the great spiritual traditions are rife with evidence that, if certain prescriptions are followed, one's godlike, universal, nonlocal nature will emerge.

But the limitations are real. One of the most troublesome is the strict way we perceive time. Margenau employs the metaphor of the "time slit" to emphasize our ability to see only a tiny slice of the entire panorama of time. Just as we can see only a narrow band in the entire electromagnetic spectrum, which we call "light," we similarly can sense only a tiny slice in time, which we call "the now." This limitation in knowing the whole of time contributes to our sense of being trapped and stranded in time—being limited to a single lifetime and feeling hopelessly mortal, destined for death.

Another major restriction on using our minds universally and nonlocally is what Margenau calls the "personal wall." The personal wall "produces the prevailing sense of individual isolation and gives us an identity as well as an ego." At its worst, it creates a sense of isolation and aloneness, which can be utterly oppressive and morbid, even fatal.

But as we shall see, neither the time slit nor the personal wall are absolute. Under certain circumstances, many of which we can learn to control, they can become "more or less opaque."

The Stochastic Wall

In addition to the time slit and the personal wall, which screen us from our identity with the Universal Mind, there is yet another impediment that crucially shapes the character of our human state: the "stochastic wall." The word "stochastic" is derived from the Greek *stochos,* which refers to "target," "aim," or "guess." This term suggests that there is randomness and uncertainty in the human condition. And who can deny it? No one really lives his life as if it were fixed and determined, not even those persons who profess a belief in determinism.

The reason, Margenau suggests, that our lives feel as if they are permeated with uncertainty is because the world at the invisible, silent, subatomic level *is* uncertain. But this is not a forlorn situation—in fact, it is just the opposite. For it is just the uncertainty of the world that allows for the possibility, at least, of human freedom of the will. But in addition to the uncertainty, another element is needed, which is choice. Thus two elements are required, Margenau states, to make human freedom a reality: choice, acting on chance.

For Margenau, there are no limitations to the highest level of Mind:

> The Universal Mind has no time slit, no personal wall; its knowledge is not limited by quantum probabilities. [25] . . . the Universal Mind has no need for memory, since all things and processes—past, present, and future—are open to its grasp. [26]

Margenau's metaphor of the time slit is rich with implications for human knowing, such as the faculty of memory. The wider the slit, the less limited in time we are, and the more we remember. If the edges of the time slit are sharp, our memory is acutely cut off at certain points in time. If the edges of the time slit are not sharp but "fuzzy," our memory will be correspondingly inexact. All the known problems with memory can be metaphorically thought of as being produced by variations in the sharpness of the edges of the time slit or by fluctuations in its width. And for all these memory problems there is a cure—the return of the individual mind, with its time slit, to the Universal Mind, which has none.

Not only the time slit, but the personal wall is also a culprit in many human maladies. In certain states of schizophrenia the personal wall dissolves dramatically so that one cannot distinguish oneself from other persons or other things. In addition, the time slit may also dilate so that one's sense of past, present, and future is demolished. The stochastic wall may also crumble, and the sense of choice and freedom may become distorted. One may feel that one has total control of all events, which may be manifested in messianic hallucinations or that one is literally God incarnate. Or the stochastic wall may become hypertrophied, made thicker and taller, so that one feels utterly paralyzed, unable to choose or act in even the simplest ways.

Often, however, these limitations may vary in the lives of quite ordinary persons, not just schizophrenics. For instance, the time slit may dilate so that the leading edge expands, allowing for precognition or "knowing into the future." Or the personal wall may lower so that one experiences a healthy empathy or connectedness with other persons and things. As Margenau describes this positive process,

> [Lowering the personal wall would] enhance our identity with others. This lowering of the wall might occur in cases of unusual sympathy with and love of others, in spontaneous empathy through concentrated attention, in meditations, in dreams, in personal experiences that . . . reveal alternate realities. It might occur in prayer, when an individual merges with the Universal Mind. The lowering of the personal wall might permit extrasensory perception in the form of coalescence of information, perhaps in the form of mind reading. [27]

Thus it is wrong to emphasize only the negative nature of the fluctuations of the time slit and the personal and stochastic walls, for many persons find these states to be genuinely uplifting and fulfilling.

Ways of bringing about these fluctuations intentionally have been known for millennia. The world's great spiritual traditions provide prescriptions which, if followed, radically modify the width of the time slit and the rigidity and height of the personal wall. They thereby teach us how to realize the eternal and infinite aspects of our being—to come to know God, the Tao, the Absolute, the One. But if one is *not* on such a carefully monitored and time-tested spiritual path, the sudden widening of the time slit and the breakdown of the personal wall can be disastrous. The unexpected confrontation with nonlocal reality can be utterly shocking and overwhelming. Perhaps the most tumultuous expression of this experience is through the use of mind-altering drugs. Here the time slit can be ripped wide and the personal wall can be demolished in a matter of seconds or minutes. Depending on many complex factors, one may describe this experience as ecstasy, higher awareness, or stark terror. A few persons have even resorted to suicide on making sudden and unexpected contact with nonlocal reality in drug experiences. Thus the decision to touch this part of one's self should never be taken lightly. It should always be pursued with a spirit of reverence as befitting a search for Truth and never, *never,* for the sake of recreation. As the noted mythologist Joseph Campbell put it, "The difference is that the one who cracks up is drowning in the water in which the mystic swims. You have to be prepared for this experience." [28]

Margenau thus recommends that the time slit and personal wall be accorded respect, for "there is a profound sense in which these are blessings within a finite human existence." [29] Even though they limit our awareness, they nonetheless help keep us intact until we are ready to undertake the search for who we really are.

Eventually, however, the time slit and the personal and

stochastic walls *must* be widened and lowered if we are to realize our universal, nonlocal nature, and the spiritual implications of a relaxation of these restrictions cannot be overemphasized. Total eradication of the personal wall and widening of the time slit to infinity would permit the merger of the person with the One, with God. Margenau describes what this might be like:

> What I . . . imply is that the conscious self will return to its presumed origin, which is the Universal Mind, and from this it seems to follow that, as part of God, it has the faculty of revisiting all aspects of its earthly experience, and perhaps even the choice of forgetting them all and consigning itself to oblivion (or even extinction). But the crucial thought, the expectation of a reunion with God, already contains some solace and hope and the promise of death as a unique experience. [30]

In summary, Margenau's vision is in the tradition of physicists Schrödinger and Bohm. The conflict between science and mankind's eternal spiritual quest is because science has not been carried far enough. If our interpretations of the physical world stop short, as in the classical world view, we see ourselves in a particle picture drifting to an end in Time. But if instead we follow the implications of the modern view of the universe, we may yet affirm the perennial perceptions of our greatest visionaries—that we are eternal, infinite, and One.

David Bohm: Group Mind in a Holographic Universe

> *Ultimately, the entire universe (with all its "particles," including those constituting human beings, their laboratories, observing instruments, etc.) has*

to be understood as a single undivided whole, in which analysis into separately and independently existent parts has no fundamental status.

—DAVID BOHM [31]

The part contains the whole; examples of this fact surround us:

Chinese boxes, each a replica of the box enclosing it, each containing an exact replica of the box enclosing it, and each containing an exact miniature of itself.

Mirrors facing each other, reflecting an unending series of identical images that gradually diminish in size, eventually fading beyond our optical acuity.

A giant oak tree producing an acorn that contains all the information to replicate itself, the succeeding oak tree continuing in the same pattern of producing its own acorns to reproduce itself, on and on.

The pattern of each human being written into the genes of sperm and ova—miniaturized and concentrated information encased in the part, yet sufficient to reconstitute the whole.

The idea that the part contains the whole is ancient, but in the modern era it has been given scientific legitimacy. Since the bespectacled Austrian monk Gregor Mendel established the notion of predictable patterns of inheritance, we have come to accept as scientific orthodoxy that wholes *are* embodied in parts. Mendel worked with peas in a monastery garden, and demonstrated that color patterns were transmitted in specific ratios with predictable regularity. His work provided the foundation for the modern science of genetics. Since Mendel, however, the scene has changed. The monk has become the physicist. Peas have given way to the universe as the object of scrutiny, and Mendel's vegetable garden has become the cosmos itself. His simple numerical ratios have become transformed into the complex mathematics of quantum theory and relativity. His description of predictable, deterministic patterns of inheritance has yielded to the language of probability and statistics.

Certain quantum physicists have expanded the principles underlying the Mendelian discoveries to an enormous scale: not only, they say, does the gene of Mendel's garden peas contain the information sufficient to reproduce the pea, each part of the universe contains all the information present in the entire cosmos itself!

This assertion is so audacious that it would be dismissed out of hand were it not for the scientific stature of its chief proponent, David Bohm. A former associate of Einstein, Bohm is now a retired Professor of Theoretical Physics at Birkbeck College of the University of London, and is one of the preeminent theoretical physicists alive today.

Bohm maintains that the information of the entire universe is contained in each of its parts. There is, he says, a stunning example of this principle in photography: the hologram (literally, "whole message"). A hologram is a specially constructed image which, when illuminated by a laser beam, seems eerily suspended in three-dimensional space. The most incredible feature of a hologram is that any *piece* of it, if illuminated with coherent light, provides an image of the *entire* hologram. The information of the whole is contained in the part. This principle, says Bohm, extends to the universe at large.

Since Bohm frequently resorts to the holographic analogy, a brief description of the process will be given. The mathematical theory underlying holograms was developed initially in the 1940s by Nobel physicist Dennis Gabor. When Gabor initially proposed the possibility, holograms could not actually be constructed—this had to await the invention of the laser twenty years later.

Holograms are created by a kind of lensless photography. Coherent light—light with waves of approximately the same frequency, traveling in phase and in the same direction—is required. This light, such as from a laser, is passed through a half-silvered mirror that allows a portion of the light to pass directly to a photographic plate. This is the reference beam. The mirror reflects the remaining portion of the light toward the object to be holographed. The object reflects light toward the recording plate, but its irregularities of surface, shape, and color disturb the coherence of the light. Thus, when the object beam reaches the plate its vibration pattern no longer matches that of the reference beam. The wave fronts of the two beams interact or interfere with one another, and the composite interference pattern is what the photographic plate records as a hologram.

Now the truly unique feature of holograms emerges. Unlike a photographic negative or slide, no image is visible on the developed plate. But when a beam of coherent light is passed through the plate, an observer on the opposite side of the plate sees a striking three-dimensional "picture" of the original object suspended in space. And if any piece of the hologram is illuminated with coherent light, the same phenomenon occurs. The smaller the piece the fuzzier the resulting image, and the larger the portion the more detailed the image becomes; but the entire representation of the original object is contained in each portion of the hologram.

Bohm proposes that the universe is constructed on the same principles as the hologram, and supports his theory by concepts from modern physics. In the modern physical view the world is not assembled from individual bits, but is seen as an indivisible whole of pattern, process, and interrelatedness. The aspect of the world that we ordinarily perceive *is* that of isolated parts, however. To us, things do seem disconnected and unrelated. Yet for Bohm this is an illusion and a distortion of the underlying behind-the-scenes oneness and unity that is an intrinsic quality of the world.

This unity, says Bohm, is "enfolded" into the universe. It is an expression of an implicit order—or, as Bohm says, an "implicate" order of electromagnetic waves, sound waves, electron beams, and numerous other forms of movement. Bohm calls this the "holomovement."

Scientists, of course, select certain facets of the holomovement for study: electrons, photons, sound, and so on,

> . . . but, more generally, all forms of the holomovement merge and are inseparable. Thus, in its totality, the holomovement is not limited in any specific way at all. It is not required to conform to any particular measure. Thus, the holomovement is undefinable and immeasurable. [32]

In order to illustrate how order can be hidden or enfolded, unapparent to the eye, Bohm uses a simple example. Imagine two concentric glass cylinders with a viscous fluid such as glycerin in the space between them. This apparatus can be rotated mechanically very slowly so that no diffusion in the glycerin occurs. Suppose you put a droplet of insoluble black ink into the glycerin and begin to rotate the system very slowly. Gradually the black droplet would be drawn out into a thin thread, eventually becoming invisible. Then if you begin to rotate the apparatus in the reverse direction, the droplet of black ink would gradually reconstitute itself, becoming visible again from the invisible black thread. The droplet of ink first became *enfolded,* invisible to the naked eye. It was not part of the *unfolded* reality that we could recognize. Yet it was still present in an *implicate* sense, and reversing the direction of rotation of the cylinder of glycerin rendered it *explicate,* visible to our senses.

Going further, Bohm suggests that holograms may be ubiquitous in nature. Although they are constructed artificially from interfering wave fronts of light impinging on a photographic plate, it is possible that this general phenomenon could be recorded in other ways. After all, light is only one expression of wave phenomena. Waves are actually commonplace in nature, and Bohm's holomovement is alive with many types of them. Electron beams could make holograms, as could sound waves, or "any form of movement," including "movements known and unknown." The universe is permeated with wave forms; and it may be, implies Bohm, that we live in a holographic universe: the *holoverse*.

The central feature of Bohm's holoverse is a oneness that exists beyond the visible world in the implicate order. It is a world we can never really know fully: We can *apprehend* it, Bohm states, but we can never *comprehend* it. It is the explicate order, the visible world of things and events, where we are most conscious; and this is a world of manifestations. It is the nature of our minds to view this outer world as real, and to take as valid the separateness we see here.

Included as part of this reality is the separateness in *space* that we feel between our self and others, which enforces the sense that we are separate minds in isolated bodies. The sense of separateness erupts also in the way we divide *time* into compartments—past, present, future—confining our individual self to only one of them, the present. But these separations are not fundamental. Instead, our world is a "single structure of indivisible links" in which all the parts, even minds, are united. As Bohm confidently states:

> Deep down the consciousness of mankind is one. This is a virtual certainty because even in the vacuum matter is one; and if we don't see this it's because we are blinding ourselves to it. [33] If we don't establish these absolute boundaries between minds, then . . . it's possible they could . . . unite as one mind. [34]

Bohm's views go beyond the unification of consciousness; they also lead to the conclusion of the immortality of the mind.

The "home" of the mind, as of all things, is the implicate order. At this level, which is the fundamental plenum for the entire manifest universe, there is no linear time. The implicate domain is atemporal; moments are not strung together serially like beads on a string. Thus Bohm concludes:

> Ultimately all the moments are really one, . . . therefore now is eternity. . . . everything, including me, is dying every moment into eternity and being born again. [35]

The similarity between the views of Bohm and those of Schrödinger (presented in chapter 5) is striking. Schrödinger also proclaimed that life is an atemporal unfolding into the present moment. In perfect accord with Bohm he said, "Every day she [Mother Earth] is bringing you forth, not *once,* but thousands upon thousands of times, just as every day she engulfs you a thousand times over. For eternally and always there is only now, one and the same now; the present is the only thing that has no end." [36]

Bohm's theories about the unification of consciousness are part of a tradition within modern physics that includes, as we have seen, some of the most highly acclaimed scientists of our era. Their theories provide support for the idea of a genuinely nonlocal mind—mind that is not limited by space and time, mind that is not confined to brains or bodies, mind that is ultimately One instead of single and individual, and mind that is immortal.

Today many people assume that the verdict from science is in: There is nothing transcendent or "higher" within the human psyche. Therefore it is useless to speak of recovering the soul because the soul does not exist. That is why the theories of these physicists are so vastly important for us. They tell us that the verdict from science on "something higher" is *not* final, that the opinion of great scientists is *not* unanimously negative, and that reasons can still be given from *within* science for the existence of the soul and its affinity with God. Even in an age of science in which God has frequently been pronounced dead, the recovery of the soul remains a project with bright hopes of succeeding.

7

A New Kind of World

What is it then between us?
What is the count of the scores or hundreds of years between us? . . .

Whatever it is, it avails not—distance avails not, and place avails not.

—WALT WHITMAN
"Crossing Brooklyn Ferry"

The Nonlocal Universe

To fly as fast as thought, to anywhere that is,
. . . you must begin by knowing that you
have already arrived.

—Chiang's advice to Jonathan
Jonathan Livingston Seagull [1]

No matter where you go, there you are.

—BUMPER STICKER

The way up is the way down; they are the same.

—HERACLITUS

What kind of world would be required to support a mind that is nonlocal—the One Mind of Schrödinger, the Universal Mind of Margenau, the holographic mind of Bohm? The answer may be, "Just the kind of world we have—a world that is itself nonlocal."

Until recently it would have been scientific heresy to propose that the world is inherently nonlocal, that an invisible connectivity unites all things no matter how disparate. Yet much evidence suggests that the universe must be understood in a new way, a way that defies the strictly local features of reality that have dominated physics since Newton.

Today most physicists believe that the world is explainable by entirely local connections, just as Newton believed, and that all currently known interactions can be explained by just four fundamental forces—the strong and weak nuclear forces, the electromagnetic force, and the gravitational force. All these forces behave as if they are mediated by fields, which have been shown to have no clear distinctions from the particles themselves. One of the features of a local reality is that the strength of forces decreases the farther one moves from their source. This is true for the "big four" forces that hold the world together. Gravity and electromagnetism diminish in strength according to the inverse of the square of the distance of separation between the source and the measuring device; the strong and the weak nuclear forces diminish even faster. All the forces that physicists focus on, therefore, seem understandable from a strictly local point of view.

Modern physics places another limitation on the idea of nonlocality through the speed of light. According to Einstein's special theory of relativity, the speed with which information can travel is limited to the speed of light or less. This is a hallowed principle in physics, and many physicists believe it is inviolable. But in a nonlocal world, connections between distant "objects" would take place faster than the speed of light: They would occur instantly.

In contrast to such a local world, what would a nonlocal world look like? A clear description is contained in physicist Nick Herbert's excellent book *Quantum Reality*. [2] Herbert, a leading authority on nonlocal effects in physics, maintains that nonlocal influences, if they exist, are not mediated by fields or by anything else. When A connects to B nonlocally, nothing crosses the intervening space, hence no amount of interposed matter can shield this interaction. In addition, nonlocal influences do not diminish with distance. Unlike local effects, they are as potent at a million miles as at a millimeter. Nonlocal influences also act instantaneously. The speed of their transmission is not limited by the velocity of light. Thus a nonlocal interaction links up one location with another without crossing space, without decay, and without delay. A nonlocal interaction is, in short, *unmediated, unmitigated,* and *immediate*. [3]

But since many physicists maintain that such a nonlocal world is impossible even in principle, and if the four fundamental forces of nature can adequately be described without resorting to nonlocal connections, why bother to talk about such a world? The reason is that in 1964 an Irish physicist, John Stewart Bell, demonstrated that such a world does in fact exist. In a mathematical proof that has since come to be called Bell's theorem and which has been confirmed by numerous experiments, he showed that the assumption that the world at heart is local is wrong. The situation that has resulted in physics in the wake of Bell's work is one of considerable tension among those physicists who bother to pay attention to it, and is described by Herbert:

> Despite physicists' traditional rejection of non-local interaction, despite the fact that all known forces are incontestably local, despite Einstein's prohibition against superluminal [faster-than-light] connections, . . . Bell maintains that the world is filled with innumerable non-local influences. Furthermore, these unmediated connections are present not only in rare and exotic circumstances, but underlie all the events of everyday life. Non-local connections are ubiquitous because reality itself is non-local. [4]

In the past twenty years Bell's theorem has been proved in various ways, most of which rely on the behavior of subatomic phenomena such as photons. But Herbert has offered a proof which shows that the theorem is applicable not only to the invisible world of quantum events but to "the familiar world of cats and bathtubs" as well. He thus contends that an essential connectedness and unity underlie all the levels of reality, not just the level of the very small.

Many physicists regard Bell's theorem as an interesting curiosity unnecessary for current work in physics. In fact, most physicists believe we are closer than ever before to a grand unified theory that will bring together the four fundamental forces in an elegant, powerful description of atomic reality. But Bell has shown that all local theories, however well they may seem to work, leave something out. As Herbert puts it, "Bell does not merely suggest or hint that reality is non-local, he actually proves it, invoking the clarity and power of mathematical reasoning. This compulsory feature of Bell's proof particularly irks physicists whose taste in realities is strictly local."

But how *could* Bell be right? The world around us is so obviously a local one. Events take time to happen at our level of reality, and we *are* surrounded by separate objects. We don't live in a world of instant happenings. Signals do have to travel from A to B, and that takes time. Yet in spite of the messages of common sense, Bell and the experimentalists who have come after him have proved that this local world of ours must rest on a deeper reality that is nonlocal. Whatever model of reality we wind up with in physics, "*it must be non-local.* . . . No local reality can explain the type of world we live in. . . . although the world's phenomena seem strictly local, the reality beneath this phenomenal surface must be superluminal [faster than light]. The world's deep reality is maintained by an invisible quantum connection whose ubiquitous influence is unmediated, unmitigated, and immediate." [5]

Some physicists argue that Bell's theorem is a sort of nuisance and a bother that will some day be swept away as more sophisticated theories of reality are developed. After all, physics theories are not eternal. But, Herbert argues, "When quantum theory joins the ranks of phlogiston, caloric, and the luminiferous ether in the physics junkyard, Bell's theorem will still be valid. Because it's based on facts, . . . it is independent of whether quantum theory is correct or not." [6]

Many of the models of reality we have examined so far are harmonious with Bell's theorem. Physicist David Bohm's theory of reality, which includes the unseen implicate order, is a thoroughly nonlocal model. So is biologist Rupert Sheldrake's theory of morphogenetic fields. So, too, is physicist Henry Margenau's model of Universal Mind, which he takes to be completely nonmaterial and nonlocal in space and time. In addition, physicist Erwin Schrödinger's theory of the One Mind is decidedly nonlocal as well. All these theories speak of a world behind the scenes that is beyond the world of objects and persons. Underlying the phenomena—such as separate, individual minds—there is a fundamental unity that is primary. Bell's theorem is important because it provides a potential validation for the concept of an underlying unity in the form of mathematical proofs and experimental data.

Herbert states,

> Bell's theorem *requires* our quantum knowledge to be non-local, instantly linked to everything it has previously touched. [7]

This is essentially a definition of nonlocal mind: mind that is linked to all else, mind that is linked to all other moments and places and persons. Yet it has not been possible to show that meaningful messages can be transmitted nonlocally between separated entities, such as two widely separated computers. Despite much effort, no superluminal signaling device has ever been built, although the nonlocal connections are there. Bell's theorem does not tell us how to use the connections among entities. This stark fact has led Herbert to suggest that perhaps these connections are not there for us to "use." Perhaps they are not "for" anything; maybe they simply are. [8]

But no matter that the theory may not implicitly allow for any direct kind of communication to take place, we nonetheless do seem to communicate nonlocally on occasions. The fact is that there are many types of nonlocal human communication. No actual contact is necessary between the communicating parties, and no signal seems to be occurring between them. Sometimes the communication seems truly to be unmediated, unmitigated, and immediate—the three main features of nonlocal reality. As we have seen elsewhere in this book, the effects of prayer and psychic healing seem to be an example of this kind of communication. And some of the most spectacular demonstrations of apparently nonlocal communication have been in the field of parapsychology, most notably by physicists Harold Puthoff and Russell Targ at the Stanford Research Institute [9] and by Robert G. Jahn, Dean Emeritus of the School of Engineering at Princeton University, and his colleague Brenda J. Dunne.

The work of Jahn and Dunne is described in their book, *Margins of Reality*. [10] It is particularly worthy because it so convincingly shatters at the everyday level the assumption of a local reality. The authors show that the outcomes of a variety of experiments cannot be explained by minds that act locally—minds limited to the present and confined to single brains. Particularly intriguing are their remote perception experiments, which have been replicated by other independent laboratories in this country. In these experiments, a stationary "sender" tries to transmit some message to a "receiver" who is at some distance, up to 6,000 miles away, and the information that is received is computer-scored. Not only has it proved possible to transmit information in ways which show that the extent of spatial separation is irrelevant, sometimes the receiver "gets" the information up to three days *before it is sent*. In a local world these events would not be possible. These experiments point to the kind of world that is foreshadowed in Bell's theorem, a nonlocal world in which changes can be transmitted in ways that are unmediated, unmitigated, and immediate—and faster even than "immediate": even *before* something has happened. Since the information was obtained *before* it was even known by the sender, it was as if the mind of the receiver could scan time into the future and know what lay ahead before it had come to pass in the present moment. In sum, these experiments suggest that the minds of receiver and sender were not really separate, but one.

Jahn and Dunne have also demonstrated decisively that human subjects can mentally influence the output of machines keyed to microscopic events that are inherently random, such as radioactive decay. In addition, this ability extends to the macroscopic domain: Subjects can influence a large-scale, random process, such as the patterns formed by randomly falling polystyrene balls in a series of open vertical slots. Interestingly, the performance of individual subjects in these tasks, over hundreds of trials, leaves a characteristic pattern of results that is different for each person—a kind of "psychic signature" that changes little from one experiment to the next.

Just as Bell's theorem seems secure because it rests on experimental facts, so too does the model of unified minds—minds that transcend space, time, and individual persons. This model, too, rests on facts. Even if quantum theory is replaced by some other theory, and if our theories of psychology and the mind are replaced by others, these facts will remain. They tell us that the world is nonlocal and that, if we look closely enough, we can see evidence of this nonlocality shining through in our everyday lives.

If the evidence for nonlocality were so subtle that it had to be teased into visibility in complex laboratory experiments such as those we have mentioned, this would be a meager harvest. We need more than differential equations, hard data, and statistical analyses to convince us of the importance of nonlocal connections in our lives. Science may show us that such things are possible, but we need to *feel* them, *experience* them, to know they are real and important.

Is there something beyond the cold, empirical demonstrations of nonlocality in physics that can satisfy the heart as well as the head? Something that might feed our spiritual as well as our scientific hunger? Many respected scientists believe the answer may be yes. Physicists J. D. Barrow and F. J. Tipler have speculated on why, if the observer in modern physics is crucial to bringing certain qualities of the world into existence, the world appears coherent among billions of observers. Why not a different world for each mind? They propose there may be "some Ultimate Observer who is in the end responsible for coordinating the separate observations of the lesser observers and is thus responsible for bringing the entire Universe into existence"—an argument, in short, for God. [11] And physicist Freeman Dyson has speculated that we may be one with this God. "I do not make any clear distinction between mind and God," Dyson relates. "God is what mind becomes when it has passed beyond the scale of comprehension. God may be considered to be either a world-soul or a collection of world-souls. We are the chief inlets of God on this planet at the present stage of its development." [12]

In Chapter 9 we will explore the spiritual implications of the modern physical view. We will see that the actual sense of connectedness, of oneness and unity, has erupted spectacularly in the spiritual lives of humans throughout recorded history, and continues to do so. We will see that the sense of oneness with other minds and with the Mind of God has left an indelible imprint on human culture in many epochs. In our search, new views of life, death, and immortality will arise. We will see that the nonlocal vision can dramatically change lives—even our own.

8

Mind and Form

Though one, He roams in various forms.

—VEDIC APHORISM

To be in any form, what is that?

—WALT WHITMAN
"Song of Myself"

In vain you believe, oh artist,
You are the creator of things.
For ages they roved in the air
Invisible for our eyes.

—A.K. TOLSTOY [1]

Mind, rather than emerging as a late outgrowth in the evolution of life, has existed always. . . , the source and condition of physical reality.

—GEORGE WALD
Nobel laureate, biology [2]

The Problem of Form

Form: where does it come from? What accounts for the shape of things?

In earlier days these questions evoked awe and wonder. Lately, however, much of the mystery of form has vanished. Molecular biologists assure us that the forms of living things are controlled by DNA, the master molecules that contain the information from which the entire organism can be assembled. Since its discovery by James Watson and Francis Crick, DNA has been regarded as the complete blueprint for living organisms.

What, then, controls the shapes of nonliving things such as crystals, rocks, or clay? They have no DNA, so other factors must be involved such as the atomic forces within the molecule describable by modern physics. In quartz crystals inner subatomic forces cause certain bond angles to form within and between the constituent atoms and molecules. These forms contribute not only to the inner configurations, but also to the external shape of the crystal as well. *All* molecules, in living and nonliving things, are configured by these forces.

Inner configurations, perhaps more than outer, account for how substances "behave" in the world. For example, it isn't enough for an amino acid to be composed of a certain number of nitrogen, hydrogen, and oxygen atoms. The amino acid molecules must have the right shape, or else the body cannot use them. Drug molecules must be shaped just so, too, or else they may be totally ineffective. L-dopa, for example, a drug that has been a breakthrough for many patients with Parkinson's disease, is one such molecule. The D-shaped molecule has an identical chemical formula, meaning that it is made up of the same atoms in the same percentages as the L-shaped molecule. However, it is ineffective because of its different shape. As a consequence, of its form, it cannot fit the receptor sites on the tissue on which it needs to act. Like a bent key, the D-shaped molecule cannot turn the lock to set certain physical processes in motion.

Life simply cannot be understood without taking form into account. What molds a leaf, the entire tree on which it grows, the shape of the caterpillar that feeds on the leaf, the birds that nest in its foliage? Plato suggested that there are ideal forms that inhabit a world of perfection from which the objects in the visible world take their shapes. This ideal world is invisible but it is omnipotent, molding the eventual appearance of all things like an invisible sculptor.

Scientists who work in the microscopic world—a biologist, say, who spends his life studying the chemical neurotransmitters in the nervous system of a particular species of snail—are generally unconcerned with the shapes of the organisms with which they work. What is important to them is the chemistry and physiology involved, which can be understood not by focusing on external shapes but by digging deeper and deeper, through painstaking analysis. On the other hand, scientists who work with whole organisms are concerned less with dissecting and have consequently seemed more susceptible to the nagging questions about form: Where does it come from and what controls it?

One such person was Hans Driesch, a German biologist who made fundamental contributions to the field of embryology in the late ninteenth century. Driesch concluded from his studies on sea urchins that the forms of living organisms, as well as their regeneration and regulation, are influenced by a nonphysical factor he called entelechy. Driesch and his entelechy, however, were vilified, for science does not look kindly on nonphysical, ghostlike entities. According to a hallowed tenet of Western thought, a cause must resemble its effect. Therefore there is no way in which an immaterial entelechy could possibly exert any influence over a physical thing and influence its form. The idea of entelechies, therefore, never caught hold in science: It simply had no place for them.

But in spite of the antipathy many scientists feel for entelechies, guiding forces, or ideal and eternal forms, the problem of form has remained a central problem in biology. No one knows why, for example, one cell matures into a leaf cell and another into a stalk cell, when both belong to the same plant and have identical DNA; or why in a particular human being one cell becomes a skin cell and another a liver cell when the DNA of each is the same. The orthodox explanation is that the complex interactions between cells, controlled and regulated by a DNA code, must be the secret, and that some day this secret will be understood.

Rupert Sheldrake and Morphogenesis

In the early 1980s a young plant physiologist from England, Rupert Sheldrake, entered this debate with a theory that shook the scientific world. His ''hypothesis of formative causation'' appeared in his book, *A New Science of Life*. [3] The theory has been the subject of passionate responses from both its supporters and detractors. It has created for Sheldrake much of the same response that Driesch's proposal of entelechy evoked: vilification and rejection by some of his colleagues in science. Still others have offered strong support, seeing it as a serious proposal that will either stand or fall on the basis of the experiments that can be made to test it. If proved correct it will rank as one of the great ideas of this century, indeed of the entire scientific era.

Sheldrake's hypothesis is relevant to our idea of nonlocal mind, with which it is most compatible. His idea of formative causation suggests that the human mind is nonlocal in both space and time; that it is not confined to the here-and-now; that it is nonmaterial and nonenergetic, implying that its effect is not diminished by spatial separation. Mind is nonlocal in another important way in his hypothesis: It is neither confined to the brain nor produced by it, although it may act *through* the brain, much as electricity acts through a wire without being generated by the wire itself. Sheldrake's hypothesis supports a collective view of consciousness wherein it may be cumulatively pooled as one mind, escaping its confinement to the brain and body of single persons. Because this hypothesis suggests that mind is nonlocal in space, that it may persist through time, and that it is beyond the body, it contains stunning implications for immortality. Moreover, consciousness in Sheldrake's hypothesis is not necessarily limited to human beings. It may be shared between many life forms, not just between humans, at least in varying degrees.

The Man

Rupert Sheldrake studied natural sciences at Cambridge University's Clare College, then spent a year at Harvard studying philosophy and the history of science. He then returned to Cambridge where he took a Ph.D. in biochemistry and cell biology in 1967, and from 1967 to 1973 he was a fellow of Clare College and Director of Studies in Biochemistry. While at Cambridge he was the Rosenheim Fellow of the Royal Society, where he focused on the development of plants and the aging of cells. In 1974 he went to India out of a desire to apply his scientific training for the aid of mankind, where he served until 1978 with the International Crop Research Institute for the Semi-Arid Tropics at Hyderabad. There he worked on the physiology of tropical legume plants. For many years he continued to spend about four months annually in India as a Consultant Plant Physiologist of the Institute, living in humble surroundings with his Indian colleagues, seeking ways to maximize the yields of food crops, which many countries desperately need.

Yet some of Sheldrake's detractors have painted him as a rogue philosopher who really doesn't know science. Actually, Sheldrake is deeply grounded in the intimate details of biology at both the cellular and macroscopic level, and has also intentionally immersed himself in the formal study of the philosophy of science, of which few scientists can claim to have knowledge.

The Hypothesis

According to Sheldrake's hypothesis, systems are organized the way they are now because similar systems were organized that way in the past. Specifically, the characteristic forms and behavior of all chemical, physical, and biological systems that exist at the present time are guided and shaped by organizing fields which, like an invisible hand, act across space and through time. Sheldrake calls these fields *morphogenetic fields* (from the Greek *morphe,* form, and *genesis,* coming-into-being).

The morphogenetic fields of all systems exert their influence on subsequent systems by a process called *morphic resonance.* As an example, one can say that the reason a plant cell becomes a leaf cell and not a root cell is because it tunes in, as it were, through morphic resonance, with the morphogenetic fields of all previous leaves of the same kind. This process occurs for all systems, wherever they may be found in nature.

The view in biology today is, of course, nothing like this. Nonetheless, the idea of invisible fields is not new. Indeed, they are an accepted part of physics today (e.g., electromagnetic and gravitational fields). So is action at a distance (which is also a feature of Sheldrake's theory), as in the moon's gravitational pull on the Earth's oceans, which creates our tides. But although fields are especially common in physics today, they have no home in orthodox biology.

Another great difference in Sheldrake's proposal and that of orthodox science is that the latter assumes that all physical processes are guided through inviolable physical laws. These laws are eternal and exist outside time. By contrast, morphogenetic fields exist *in* time: They are built up with the passage of time, they are modified by the shapes and forms of all subsequent systems, and they sweep through time to influence future systems which are yet to be.

It is not only Sheldrake who has raised questions about the assumption that physical laws are timeless and immutable, many other scientists working within accepted frameworks have begun to cast doubts on them as well. Perhaps the greatest challenge comes from current theories in cosmology. It is widely believed that the universe originated in the unimaginable instant called the Big Bang. In this colossal moment all the matter in the universe came into being, along with all the physical laws regulating its behavior. As there was nothing material or physical before the Big Bang, it makes no sense to speak of "physical" laws existing then. And if they did not exist then, they cannot be eternal. The laws as we know them, then, are derivative and developmental, not given for all time. Seen against the backdrop of modern cosmological theory, Sheldrake's suggestion that physical laws are not changeless and eternal, but develop within time as things "go along," does not appear as heretical as it might seem.

But the cosmologists do not go as far as Sheldrake. They suggest that *after* the Big Bang the laws calcified; they became fixed and thereafter never changed. In contrast, Sheldrake claims that the laws guiding the forms of all things are always changing through time, since the morphogenetic fields are always susceptible to modification.

The Furor

The publication of *A New Science of Life* in England ignited a firestorm of controversy. Sheldrake was compared to Uri Geller, the psychic who is well known for his spoon-bending ability, and was accused of "going mystic."

Other scientists rose up to speak in defense of both Sheldrake's ideas and the process of open inquiry, stating that no ideas should be condemned before they are tested, no matter how outrageous they might seem. The physicist Brian Josephson, a Nobel laureate, stated,

> A new kind of understanding of nature is now emerging, with concepts like implicate order and subject-dependent reality (and now, perhaps, formative causation). [4]

Among the strongest supporters of Sheldrake's hypothesis of formative causation in the early days following its publication was the English magazine *New Scientist*. This publication boldly stated that "Western science has sadly misconstrued the world and all the creatures therein. . . . What Sheldrake proposes is scientific. This does not mean he is right, but that it is testable." [5]

One of the first people in the United States to recognize the potential importance of Sheldrake's ideas was Marilyn Ferguson, editor-publisher of *Brain/Mind Bulletin* and author of the highly acclaimed book, *The Aquarian Conspiracy: Personal and Social Transformation in the 1980s*. [6] As Ferguson saw it,

> [Sheldrake's] startlingly new hypothesis . . . could overturn many of our fundamental concepts about nature and consciousness. In its implications, it is as far-reaching as Darwin's theory of evolution. [7]

In living things, all processes of development start from systems that already have their own characteristic patterns of organization. The embryo developing inside a fertilized egg, for example, contains nucleic acids and proteins that are organized in a specific way. Not only is there a specific chemistry involved in these substances—certain formulas that describe the molecular content of nitrogen, carbon, hydrogen, oxygen, and other elements—there are specific forms that have already begun to take shape inside these molecules themselves.

Form, therefore, begins at the earliest stages in living organisms, starting in the innermost molecular recesses. And the forms contained within the molecules are shaped by morphogenetic fields, Sheldrake says, that have already been set in place by similar molecules of past systems.

Many persons visualize morphogenetic fields as some invisible sculptor working from the outside of systems, hammering or carving them into certain shapes—molding an arm here and a feather there, making an elephant into an elephant and a salmon into a salmon through some benevolent, skillful means. Although this image might serve as a convenient shorthand for Sheldrake's views, it is possible to view the work as being done from the inside, rather than from the outside, because the morphogenetic fields function as *patterned restrictions* on the multitude of probabilistic, indeterminate events occurring at the deepest levels of physical systems. It is there, in the inner shapes that atoms and molecules assume, that morphogenetic fields may be felt first. This process may then sweep outward, manifesting itself eventually in the visible, external shapes of things. Thus the process of formative causation can be regarded as an "inside job," not an outside one.

Even emotions and thoughts may be affected by these fields, inasmuch as our feelings are affected by our internal chemistry. This is one reason why the fields are potentially so pervasive, and why we can jump from atoms to thoughts in discussing their activities. In fact, there is no "jump" at all, once we see how the fields are thought to work—where their effects begin and where they end.

But there really is no "end" to the expression of these fields. For once any individual person or any other living thing is affected by them, they in turn affect the fields themselves by cumulatively adding to them. All things, thoughts, and behaviors are plowed back into the great forward sweep of morphogenetic fields. This is in stark contrast to the view that dominates current science, which says that an organism's influence in the world is terminated when it dies. According to the hypothesis of formative causation, there is an endless absorption into the Great Pool of Being, from which one is eternally brought forth thereafter, manifesting as influences on the inner and outer patterns of things newly born.

The simplest way to understand morphic resonance is through an analogy on which Sheldrake always relies, that of a TV or radio set. In the TV the wires, transistors, and other components together act as a tuning device that picks up the signal from the TV station. The eventual picture that is displayed depends on the internal elements of the set being tuned properly to the transmission. If one changes the components, one may change the tuning and interfere with the picture. This may result in not only distortion, but the loss of the picture altogether.*

By analogy, in a developing egg the DNA and the other chemicals that it contains give rise to the "tuning characteristics" of that particular species, just as the TV set can pick up a certain range of signals and no others. Analogously, the developing egg can "resonate" with certain morphic fields that have been set up by similarly developing eggs in the past. This causes the particular egg to become a chicken egg, for example, and not a partridge or eagle egg.

Similarly, the brain has its component parts—its neurons, blood vessels, supporting structures, and so on. It produces mental images, thoughts, emotions, and sets various motor events in action. But it no more brings these events into existence than the TV set produces its own picture.

When it comes to the relationship of the mind and the brain, conventional materialists deny that these analogies carry any truth. The brain is said to be the origin of consciousness. There is no external source whence the "signal," the mind, originates. As evidence, physiologists have long pointed to the evidence that damage to the brain causes certain functions to drop out—speech, hearing, and vital functions such as the heart beat or respiration. Or in surgery one can stimulate certain parts of the brain and produce actual thoughts, compulsions to act, images, spoken words, or the movement of a particular body

---

*The idea that the brain is a receiver or instrument upon which the mind or consciousness operates is ancient. Four centuries before Christ, Hippocrates stated that "the brain is the interpreter of consciousness," and, "to consciousness the brain is the messenger." (Hippocrates, "The Sacred Disease," *Hippocrates*, vol. 2, W.H.S. Jones, trans. [Cambridge, MA: Harvard University Press, The Loeb Classical Library, 1952], p. 179). In our own century, the philosopher Henri Bergson expressed the same idea, as has a small but increasing number of scientists. Among them is the eminent neurologist Sir Charles Sherrington, who said, speaking of mind and brain, "That our being should consist of *two* fundamental elements offers I suppose no greater inherent improbability than that it should rest on one only." (Sir Charles Sherrington, *The Integrative Action of the Nervous System* [New Haven: Yale University Press, 1947], p. xxiv; first published in 1906).

part—more evidence, it is said, that all our mental and motor life is in the brain. And doing away with brain function altogether erases the mind completely, it is assumed, as after prolonged anoxia or severe trauma.

But let us return to the TV analogy and press it further. Imagine a naive person who has never seen a TV before. Looking at the picture, he asks himself where it is coming from. He might judge that it originates inside the box itself—not an unreasonable assumption. He investigates the TV by taking it apart, pulling a wire here, changing a connection there. With every modification the picture changes—fuzzy images at first, then a complete loss of the picture as he progressively maims the components, leading him to conclude that, since he has disturbed the inside of the TV set and lost the picture, this is where the origin of the picture must lie. His "proof" parallels the logic of the brain mechanist who firmly believes that damage to the brain proves that that's where the mind originates. In both cases, it's all in the machine.

In both instances the reasoning overlooks the fact that external forces may be at work—the TV signal and, in the case of the brain and developing systems such as embryos, says Sheldrake, consciousness and morphogenetic fields. Everything is not present in the brain nor the DNA. Up to a certain point the brain physiologist is right: one *can* scramble the output by interfering with the wires, transistors, DNA, proteins, blood, oxygen, and other components. However, this is not the complete explanation either of the TV set, the brain, or the developing embryo.

"But then the mechanists will say," says Sheldrake, " 'well, we admit we can't explain it now, but we will be able to explain it in the future.' They issue undated promissory notes. It's essentially an act of faith in the mechanistic method, not really a strict scientific hypothesis." [8] And a recurring theme in the promissory note is the role of DNA. It carries the genetic code, which is somehow supposed to govern everything that happens to developing, living things. Bone and ear and liver cells all contain the same DNA, so there must be something over and above the DNA itself to account for why they turn out differently. The genetic code is said to be the hidden factor. But Sheldrake believes that in postulating the genetic code as some all-powerful ordering device that somehow pushes cells toward a goal—to become a white blood cell instead of a kidney cell, say—mechanists are inserting something that is suspiciously like morphogenetic fields themselves, which also have a goal. This concept of the genetic program is, after all, teleological—it presses toward a specific goal—just as the proposed morphogenetic fields do. This goes far beyond the mechanistic approach itself, which ostensibly denies any purposes or goals in nature. To this extent there is an unexpected similarity between Sheldrake's hypothesis and the ideas surrounding the action of the genes.

But the similarity does not go very deep. Sheldrake sums up the problems surrounding the DNA-dominated thinking in biology:

> DNA, by providing the code for the sequence of amino acids, enables the cell to make particular proteins. That is *all* the DNA can do. . . . But the problem with morphogenesis is not just the question of getting the right proteins in the right cells at the right time. It's how, given those proteins, the cells organize themselves into particular forms; how cells group together in tissue of particular forms; and how those are shaped into organisms of particular forms. DNA helps us to understand how you get the proteins which provide, as it were, the bricks and the mortar with which the organism is built, but it doesn't explain how these bricks and mortar assemble into particular patterns and shapes. The idea of DNA shaping the organism or programming its behavior is a quite illegitimate extrapolation from anything we know about what DNA does. . . . Everything to do with heredity and properties of living organisms is being projected

> on to DNA within the mechanistic model—all the unsolved problems of biology are being attributed to DNA. . . . So what starts as a rigorous and well-defined theory about the way DNA codes the RNA and how RNA codes the proteins, soon turns into a kind of mystical theory in which DNA has unexplained powers and properties which can't be specified in exact molecular terms in any way at all. . . . These extra things that it is supposed to do are, I think, what in fact morphogenetic fields do. [9]

There currently is great hope that the developing forms of embryos can be understood by the action of so-called morphogens, chemicals which, presumably under the direction of the DNA, control the shapes that the embryos take. Early work appears promising, and one particular chemical, retinoic acid, has already been shown to have some effect in this regard. [10] Yet it is not clear that the identification of these compounds will solve the problem of form, for the mechanist would almost certainly continue to place them under the control of DNA, since it is heretical to allow anything to escape the hegemony of the genetic code. The identification of morphogens would seem, then, merely to insert another link into the mechanist's chain and not answer the problems to which Sheldrake has pointed. And in any case, Sheldrake is concerned not only with the forms that living things such as embryos take, but the forms of nonliving things such as crystals as well. And no morphogens have been hypothesized that would explain the appearance of form in them.

According to the doctrine of extreme or radical materialism, consciousness must be discarded except as a function of the physical processes occurring in the brain. But one is not forced into this dire position under the hypothesis of formative causation. In it one can follow a commonsense view, as Sheldrake puts it, of recognizing the reality of one's own consciousness. Surely this is not demanding too much. After all, this is what all human beings—even dedicated materialists—do anyway.

Creativity and the Origin of New Forms

But what about entirely *new* forms? They are not conserving any preexisting shapes. Sheldrake leaves the question of new origins open because he sees it lying outside the scope of natural science, in the domain of metaphysics. But he does speculate about the capacity for the new and the original to occur in nature, and acknowledges the influence of the French philosopher Henri Bergson on his thinking. Bergson was noted for his belief in *élan vital,* an original life force that guided living things in their historical development. As Sheldrake states,

> Bergson . . . was very keen to assert the genuine creativity of the evolutionary process. He again and again pointed out that our minds have a tendency to deny creativity. We can't explain [it]. It involves the completely new, the original. So we prefer to say that creativity is not creativity at all but merely the expression of something ''archetypal'' that already exists in a latent form. That denies true creativity. It's like saying everything is laid down in advance, and that evolution is rather like unrolling a long carpet—it is just unrolled in time. [11]

And what about ultimate origins, the origin of the universe? Where does it come from? Sheldrake answers:

> The universe itself [does have] an origin, and both the creativity within the universe and the universe itself require an explanation. And these can only be explained in terms of something which is over and above or beyond the universe—in that sense transcendent. This would correspond to the traditional theistic views of creation, which would have a God who is beyond, above, and in nature. . . . I myself take this view. [12]

Nonlocal Mind and Sheldrake's Hypothesis: The Cosmic Memory Bank

One of the great correspondences between the hypothesis of formative causation and nonlocal mind is the two-way process linking past and present. The past in some sense *is* the present, because the present shapes the past by feeding back into it and modifying the preexistent morphogenetic fields. Every event adds the effect of its own coming-into-being to the morphogenetic field with which it resonated, or else it begins a new morphogenetic field, in either case persisting into the future.

Now, when we begin to apply psychological language and talk about thoughts instead of material events such as the development of embryos, the same process can be imagined. There is a two-way process linking present and past—past thoughts influencing present ones via the morphogenetic fields, and present thoughts adding to or modifying the fields. The present does not come into being only to die; it is preserved in an invisible morphogenetic record that thereafter makes a contribution to future events. In this way thoughts are plowed back into the universe, into a kind of cosmic memory bank, as Sheldrake puts it. It is possible to envison a kind of universal mind taking shape. This brings to mind the words of Sir James Jeans, the English astronomer-physicist:

> The concepts which now prove to be fundamental to our understanding of nature . . . seem to my mind to be structures of pure thought . . . the universe begins to look more like a great thought than like a great machine.[13]

This view of the universe suggests that it is brimming with thought, alive with mind and consciousness. This is a view that sets the stage for nonlocal mind—mind unrestrained by space and time, mind not confined to the brains and bodies of single persons.

Sheldrake has been quick to point out that many traditions have held the notion of a cosmic mind, among which is Mahayana Buddhism with its idea of *alayavijnana,* or storehouse of consciousness. And according to Theosophists there is a similar process, that of the *akashic record*. This view holds that everything that happens, whether physical or mental, is encoded in subtle dimensions of space and time, where it functions as a data bank for karma, an idea found also in Tibetan Buddhism. [14]

Does the hypothesis of formative causation rest on anything more than speculation? The theory of morphogenesis has a long background in biological science. Long histories do not count as evidence, but they at least show that the idea was not hatched yesterday and therefore might command a closer look. The fact is that Alexander Gurwitsch and Paul Weiss developed the basic concept of morphogenetic fields in the 1920s. The reknowned biologist C. H. Waddington also took up the theme, but considered them to be a mere ''descriptive convenience,'' something of a shorthand for the chemical interactions that actually go on in living things.

Much of the objection to Sheldrake's hypothesis is because it suggests that acquired characteristics can be inherited, a view that is generally attributed to the early nineteenth-century naturalist Jean Baptiste Lamarck. Lamarckism is dogmatically denied in modern biology. A plant, for example, that becomes stunted by growing in a severe environment will not produce offspring that are stunted; or if a chimpanzee learns a certain skill, it cannot be transmitted genetically to subsequent generations. The genetic code cannot be changed by such external circumstances, no matter if external characteristics such as form, appearance, skills, or knowledge of the organism change—thus says modern biology.

Perhaps the most supportive data for the existence of morphogenetic fields comes from a series of experiments begun by the psychologist William McDougall at Harvard University in 1920. His goal was to find out if animals could inherit abilities acquired by their parents. McDougall put white rats, one at a time, in a tank of water. The only means of escape was by swimming to one of two gangways and climbing up. One gangway was lit brightly and the other was dark. If the rats left by the illuminated escape path they received an electric shock, but not if they took the darkened gangway. McDougall recorded how many trials were required for the rats to learn to escape consistently by the unlighted path.

The first generation of rats required an average of over 160 shocks each before avoiding the illuminated gangway. However, the second generation, bred from these experienced parents, did much better. And their offspring did better still. McDougall bred thirty generations of rats, at which point the rats were making only twenty errors each.

McDougall concluded that these results pointed to the inheritance of acquired characteristics, but this conclusion was extremely controversial. It raised the specter of Lamarckism. McDougall's experiments were scrutinized by the leading biologists and no one could find fault with his experimental design. It was decided that somehow he must have unknowingly bred the most intelligent rats from each generation, in spite of the fact that he chose the parents at random.

McDougall took up the challenge. He began a new experiment in which he selected only the most stupid rats from each generation to be parents for the next. One would have expected, according to the tenets of orthodox genetic theory, that subsequent generations would have performed progressively poorer. Instead, they did better: After twenty-two generations the rats were learning ten times faster than the first generation of stupid ancestors.

These results were stunning. Other researchers hastened to replicate McDougall's experiments—Dr. F.A.E. Crew in Edinburgh, and Professor W. E. Agar and his colleagues in Melbourne. They made similar tanks and used white rats of the same breed. Something unexplainable happened in their experiments: From the very first generation their rats learned the task much faster than the first generations of McDougall's rats. In fact, some of the first rats Crew tested "learned" to escape through the darkened runway at once, without making a single error.

Agar not only studied the change in the rate of learning of successive generations of rats descended from trained parents, but also that of a parallel line of rats descended from untrained parents. In this control line, some of the rats were tested in the water tank and then discarded, while others that had not been tested served as the parents from which the next generation was bred. Agar's studies went on for twenty-five years. They confirmed McDougall's findings: Successive generations in the trained lines tended to learn quicker and quicker. However, so too did rats in the control line.

Since the improvement in the rates of learning occurred in successive generations of the untrained control line as well as of the trained rats, the results could not be due to the passing on of genes that might have been modified through learning—the transmission of acquired characteristics. Thus, although McDougall's conclusions were disproved, his results were confirmed. They have never been explained and still remain a curiosity, totally incongruous with conventional concepts of genetics. Yet they fit extraordinarily well with the concept of morphogenetic fields.

One of the reasons McDougall's work did not achieve more acceptance was that it smacked of Lamarckism, that acquired characteristics could be inherited. But Lamarck's ideas, after being relegated to the junkheap in biology, may be making a comeback, for recent findings suggest that evolutionary and genetic changes may be *directed* by external circumstances.

In one study, Harvard's John Cairns and his coworkers found that bacteria genetically incapable of metabolizing the sugar lactose can become lactose-consumers when placed in an environment in which lactose is their only food source. In order to consume the sugar and thus avoid starving, the bacteria had to mutate at a rate far greater than that known to occur randomly. Thus the bacteria rapidly acquired a new characteristic and passed it on to future generations, violating the current prohibitions against such. [15] In a similar experiment, researcher Barry Hall at the University of Connecticut subjected bacteria to an even more difficult task. In order to metabolize the novel sugar that was their sole nutrient, they had to perform two mutations, one of which was an excision of existing genetic instructions. The odds against the two mutations occurring randomly during the same time period are more than one in a trillion. [16] As in Cairns' experiment, the bacteria then passed on the newly acquired ability to subsequent organisms.

The significance of these experiments is currently the subject of intense debate. Are the results limited to bacteria? What percentage of all mutations are accounted for by these newly discovered mechanisms? Are most genetic variations still attributable to randomness, with "experience" playing only a minor role? Regardless of how the debate is settled, these experiments suggest, Cairns states, that all variation is not random, but that *the genetic packet of an individual cell can profit by experience*. Cairns summarizes the current situation:

> The early triumph of molecular biology strongly supported the reductionists. . . . Curiously, when we come to consider what mechanism might be the basis for the forms of mutation . . . we find that molecular biology has, in the interim, deserted the reductionist. Now almost anything seems possible. [17]

Since the hypothesis of formative causation was introduced, several experiments have been done to prove its validity. These experiments are surprisingly simple and straightforward, which is a great departure from the elaborate laboratory apparatus that usually characterizes the modern scientific approach to biological research. The studies have examined the rates of mutations in fruit flies, and pattern recognition and rates of learning in human beings during various tasks. International prizes have been given for designing experiments that might either prove or disprove the hypothesis. To date the results are equivocal. Most of the outcomes are compatible with the theory but a few are not. [18] Sheldrake, as well as many other scientists who have carefully scrutinized his hypothesis, believe that it is capable of disproof as any good proposal should be and that it will either stand or fall on the basis of the experimental evidence.

Mind-to-Mind Communication: Morphic Resonance or Telepathy?

The presence of morphogenetic fields provides a way for all thoughts to become linked across space and time. This is a picture of nonlocal and transpersonal mind, and a way for individual minds to communicate.

This fact has not gone unnoticed by the critics of the hypothesis of formative causation. Some of the earliest objections, in fact, were that it aided parapsychologists and believers in telepathy, ESP (extrasensory perception), and paranormal phenomena in general. Sheldrake agrees there may be a connection between morphic resonance and telepathy, and that, in fact, the differences in them "may just be a matter of semantics. We do not know what telepathy is, and morphic resonance may be a much more general theory of which telepathy is a special case, involving connections between specific individuals. With crystals or plant forms one ordinarily doesn't refer to telepathy. So maybe we are talking about the same thing in different ways." [19]

Experiments with animals seem to indicate that telepathy or morphic resonance (acknowledging for now the difficulty in distinguishing the two) occurs not just between humans, but between humans and animals as well. The "lost animal" anecdotes recounted in Chapter 4, in which animals manage to find their way back home or to places they've never been across enormous distances, unexplainable by known sensory cues, could be accounted for by human-to-animal telepathy. The human owner of the pet, who knows the way back home, could be "sending" the information to his pet (in the language of telepathy). Or his thoughts could be establishing a morphogenetic field that makes it possible for the knowledge of the way back home to recur in the mind of the pet (in the language of morphic resonance). In either case the information about how to return might get through.

The Oak School experiment of psychologist Robert Rosenthal of Harvard, described in Chapter 3, could be interpreted in the same way. The teachers expected the "superior" students to be achievers, and they were, perhaps because the teachers' expectations molded or shaped the behavior of the students in that particular direction. The same process could have determined the outcome of Rosenthal's earlier experiments with the rats, which was of the same design. And also with the study of Cordaro and Ison's planarial worms which, like the students and the rats, behaved as their experimenters expected them to. These studies imply not only human-to-human, but human-to-animal communication; however, it seems impossible to decide whether morphic resonance or telepathy is the better explanation. In either case the mind seems to be behaving nonlocally.

Many other studies with animals seem to show that they exercise nonlocal mental connections. Scientists have shown that dolphins in one area can suddenly develop a particular behavior pattern after dolphins in a remote area have already demonstrated it. In one experiment, researcher Wayne Doak describes an extraordinary experience involving a dream. In the dream he was told, when working with dolphins, to perform a certain significant activity and to repeat a word that in the Maori language means "the sound the dolphin makes with its blowhole." The dolphins would then perform a very specific behavior pattern, the dream revealed. The next time he was swimming with the dolphins he did as the dream told him—and the dolphins performed exactly as the dream had predicted! He was not only astonished by *this* event, but also by the fact that a friend working with dolphins 3,000 miles away, reported the following day the same experience with the same word and same behavior from his dolphins. [20]

Perhaps what is important is not deciding whether telepathy or morphic resonance is the better explanation in these interactions, but to acknowledge them as legitimate events in need of explanations. Neither telepathy nor morphogenetic fields has been explained, although science today, particularly subatomic physics, is alive and buzzing with theories of invisible fields. If our own minds also prove to be fieldlike, if they are shown to be unconfinable to point-places in space and time, and if they are demonstrated to exist beyond brains and bodies, then Sheldrake's hypothesis of morphogenetic fields may prove to be a valuable step in our understanding.

Part III

GOD: The Synthesis

We must assume our existence as broadly *as we in any way can; everything, even the unheard-of, must be possible in it. That is at bottom the only courage that is demanded of us: to have courage for the most strange, the most singular, and the most inexplicable that we may encounter. That mankind has in this sense been cowardly has done life endless harm; the experiences that are called "visions," the whole so-called "spirit-world," death, all those things that are so closely akin to us, have by daily parrying been so crowded out of life that the senses with which we would have grasped them are atrophied. To say nothing of God.*

—RAINER MARIA RILKE [1]

The spirit of man communes with Heaven; the omnipotence of Heaven resides in man. Is the distance between Heaven and man very great?

—*Discourses on Vegetable Roots* (Ming Dynasty) [2]

9

Spiritual Implications of Nonlocal Mind

We consider bibles and religions divine—I do not say they are not divine,
I say they have all grown out of you, and may grow out of you still,
It is not they who give the life, it is you who give the life,
Leaves are not more shed from the trees, or trees from the earth, than they are shed out of you.

—WALT WHITMAN
"A Song for Occupations" [3]

No one has seen God and lived.
To see God we must be nonexistent.

—HAZRAT INAYAT KHAN

I, Lord, went wandering like a strayed sheep, seeking Thee with anxious reasoning without, whilst Thou wast within me. I went round the streets and squares of the city seeking thee; and I found thee not, because in vain I sought without for him who was within myself.

—ST. AUGUSTINE

But what a thing dost thou now
Looking Godward to cry,
'I am I, thou art thou,
I am low, thou art high?'
I am thou, whom thou seekest to find him
Find thou but thyself, thou art I.

—SWINBURNE

Draw if thou canst the mystic line
Severing rightly His from thine
Which is human, which divine.

—EMERSON

Though difference be none, I am of Thee,
Not Thou, O Lord, of me;
For of the sea is verily the Wave,
Not of the Wave the Sea.

—SANKARACHARYA
8th Century

My goal in exploring the nonlocal aspects of human nature has not been to arrive at a clever theory of how things work. We have an abundance of clever theories today, and something more than cleverness is needed if we are to stop killing ourselves and devastating our planet, the only home we have. The "something more" that is needed is a vision that will transform our view of who we are and how we fit into the way of nature—a new world view that would redefine our very being. I believe the view of man as a nonlocal being offers something of the new vision we need.

The nonlocal view of man places mind and consciousness outside the person, the brain, and the body, and it leads to a theory of the One Mind, which is boundless in space and time. As I have tried to show, there are many, many reasons to take this possibility seriously. These reasons arise not only from the eternal intimations of seers and visionaries but also from modern scientific insights.

But it gets stickier. The next step in nonlocal thinking not only gets us outside the brain and the body and the individual person, it takes us beyond mankind altogether. It lands us squarely in the lap of God—or the One, Logos, Tao, Brahman, Buddha, Krishna, Allah, Mana, the Universal Spirit or Principle. Nonlocality at this point ceases to be an intellectual plaything, if it ever was. It assumes monumental proportions. It becomes theological dynamite—something sobering, like a child's toy with a nuclear warhead—the instant we realize the spiritual implications that are involved. Because nonlocal mind is *boundless* and *limitless,* and because these are precisely qualities of God also, this means that at some level we share something with the Godhead. To some degree, the borders of man and God overlap. This is a way of saying that there is an element of the divine within man, which shows itself through man's nonlocal qualities. Put another way, nonlocality is the stage on which God and man meet.

If "being God" has always had a blasphemous, irreverent, and heretical connotation in our Western religious traditions, then this underscores the time-and-spacebound qualities of our religions. In essence they are built on classical notions of how the world works. They emphasize linear time with their emphasis on becoming, on getting somewhere other than where we are at the moment. This may come through grace, good works, a combination of the two, or by some other means; but underlying this way of thinking is a local view of the world. Improvement is always needed, and improvement is always in Time. It is always a state that does not yet exist and that must be realized at some future point. And these injunctions for improvement or for salvation are always addressed to the individual person. It is the single and separate being who is in need and who must somehow become different than he happens to be at the moment. This way of thinking is clearly local, emphasizing linear time, local space, and the single person.

But this really is only part of the spiritual picture our species has developed, although it is a persistent and widespread part. Many spiritual traditions, especially those of the Orient, are much more nonlocal in flavor. And even within the Christian tradition of the West there are esoteric schools that are also of a nonlocal nature. These traditions do not emphasize the linear nature of time and the depraved nature of man, with the resultant need to become radically different through a redemptive process that works into the future. Neither do they emphasize the primacy of the individual. Instead they urge all persons to wake up to their already-present oneness with each other and with their Creator, their inherent perfection, and their intrinsic divinity. They demand enlightenment, which means becoming aware of what is *already* present, not some yet-to-be-realized redemption. And they repeatedly state that the belief in the isolated self, the ego localized in this body and person, is a fiction. These spiritual leaders have made the Great Nonlocal Leap—the most radical breakthrough in the concept of ourselves we can possibly have—a movement away from a limited to an open perception of reality and the self.

Throughout history all major cultures have produced spokespersons for this point of view. In certain instances, particularly in the West, these persons have paid a great price for enunciating their nonlocal spiritual visions. The problem frequently has been that they sound too much like heretics to remain unchallenged by the dominant spiritual view of the culture, which has almost always been of a local nature. In this section we will continually draw on the actual words of these visionaries.

The Great Nonlocal Vision: The Meeting of God and Man

One of the most dramatic examples of the call to spiritual nonlocality can be found in the *Corpus Hermeticum,* dating back at least two thousand years. In it we find these remarkable words:

> Unless you make yourself equal to God, you cannot understand God: for the like is not intelligible save to the like. Make yourself grow to a greatness beyond measure, by a bound free yourself from the body; raise yourself above all time, become Eternity; then you will understand God. Believe that nothing is impossible for you, think yourself immortal and capable of understanding all, all arts, all sciences, the nature of every living being. Mount higher than the highest height; descend lower than the lowest depth. Draw into yourself all sensations of everything created, fire and water, dry and moist, imagining that you are everywhere, on earth, in the sea, in the sky, that you are not yet born, in the maternal womb, adolescent, old, dead, beyond death. If you embrace in your thought all things at once, times, places, substances, qualities, quantities, you may understand God. [4]

Here the writer takes us beyond time to eternity, beyond single events to all events, beyond the here to the everywhere, beyond the self to unity with God. In one short paragraph he explodes the local view of reality and tells us that we *must* explode it if we want to know God.

And so he leads us to the bugbear of nonlocality: God-unity, which sets up for some people an image of spiritual hubris. But the author of the *Hermeticum* is not equating becoming "equal to God" with usurping Divine power. It is impossible to steal what one already possesses. One cannot masquerade as what one already is, only what one is not. It is impossible to "play God." Moreover, when one sees the illusory nature of the individual self, one realizes there is no one to do the usurping in the first place.

Although mystics the world over have glimpsed a harmonious vision of the unity of God and man, this realization nonetheless has still caused an immense amount of "spiritual indigestion" in all cultures and traditions, not just those of the West. Alan Watts, who did much to bring the philosophy of Zen Buddhism to the West in our century, clearly expressed this dilemma and proposed a way out:

> There has not yet existed a religion or a philosophy in which there is a true marriage of Heaven and Earth. There have been many approximations, but never one in which each said to each other, without any shadow or reservation, "I love you with all my heart!" We have seen the spiritual reduced to the material, and the material vaporized in the spiritual. We have seen unhappy compromises in which the spiritual is always saying to the material, "Yes, but. . . ." We have seen the material perpetually damned with faint praise and always being talked down for the odd reason that it constantly changes and flows away, as if that were something wrong. Only most occasionally have Hindus had the courage to swing fully with the Lord in dancing his *maya*, and to stop insinuating that in some nasty, niggling last analysis the physical universe

> shouldn't be happening. Christian theologians, too, seem to have a commission to *protect* the Deity from full union with his universe, as if this would somehow completely subvert his morals. Indeed, they allow that the creature may participate in, reflect, be adopted into, transfigured by, or given unity with his Creator. But always, at the end of the line, there is someone wagging his finger and saying, "But . . . never forget, little creature, that you are nothing, nothing right down to your miserable essence, for your being is not your own." So—my body is God's, my mind is God's, my being itself is God's, all on loan to nothing and to no one.
>
> . . . The thing is to see all faces as the masks of God, all characters as his roles, preachers included. Toward the end of his life that extraordinary Hindu-Buddhist-Muslim saint, the poet Kabir, used to look around and ask, "To whom shall I preach?" He saw the face and the activity of his Beloved in every direction.
>
> Obviously, as so many Christians seem to fear, this vision of God-as-all might be used as rationalization for indulgence in total wickedness. But fire is not untrue, or something to be abolished, because it can be used to burn people alive. [5]

Paul Brunton, a modern Western mystic who spent many years in the Orient, and whose voluminous writings are only now coming to light posthumously, had this to say about the union of God and man:

> Those who mistrust this mysterious teaching sometimes allege that its deification of self is an attempt to equate God with the human personality, and to depose Deity in order to enshrine a part of His creation. This is a misunderstanding. Whoever enters into the expe-

rience of contacting the depths of his inmost being can emerge only with deeper reverence for God. He realizes his helplessness and dependence when he thinks of that Greater Being from whom he draws the very permit to exist. Instead of deifying the personal self, he has completely humiliated it. Self, in its ordinary sense, must indeed be cast away that God shall enter in. [6]

From the West to the East, the realization has surfaced that man, when he reaches out with humility and honesty, cannot really usurp God. Speaking from the ninth century A.D., the Hindu mystic Shankara:

> As Brahman constitutes a person's Self, it is not something to be attained by that person. And even if Brahman were altogether different from a person's Self, still it would not be something to be attained, for as it is omnipresent it is part of its nature that it is everpresent to everyone. [7]

The idea of man's godhead is not limited to ancient mystics. From our own century the great physicist Erwin Schrödinger discovered the same message:

> In Christian terminology to say: "Hence I am God Almighty" sounds both blasphemous and lunatic. But please disregard these connotations. . . .
>
> In itself, the insight is not new. The earliest records . . . date back some 2500 years or more. . . . The recognition . . . [that] the personal self equals the omnipresent, all-comprehending self . . . was in Indian thought considered, far from blasphemous, to represent the quintessence of deepest insight. The striving of all the scholars of Vedanta was, after having learnt to pronounce with their lips, really to

> assimilate in their minds this grandest of all thoughts . . . that can be condensed in one phrase: DEUS FACTUS SUM (I have become God). [8]

The nonlocal vision of the universe is thus not an invitation to blasphemy. It requires an expanded temporal and spatial sense, and a broader sense of the nature of being human.

There is one point on which all the great traditions agree: God is omnipresent and eternal—that is, he is nonlocal in space and time. If we can hold this fact in our mind, perhaps we will find it easier to understand why the urge to Him has always been an urge to nonlocality.

Christianity and Nonlocal Mind

> *Christianity has lost the teachings of Jesus. . . . It has made him a special individual, with gifts we do not have. But he said, "The works that I do shall [you] do also, and greater works." He spoke of the 'local' mind in relation to the one Mind when he said: "I and my Father are one" and "My Father is greater than I"—the 'local' mind is one with the one Mind but of course the one Mind is more than the local mind. He was referring to the same idea from a different angle when he said: "Give, and it shall be given unto you"—because you are giving only to yourself, as an individualization of the one Mind. Even the Golden Rule, in its negative or positive form, repeats the point. Don't do to others what you do not want them to do to you, or do to others what you would want them to do to you—*

what you don't do or do to others is what you don't do or do to yourself, because there is one Mind!

—ROBERT DOLLING WELLS [9]

Orthodox Christianity is overwhelmingly local in its perspective. It has the imprint of the commonsense world view with its familiar space and time: It assumes a person estranged and distant from God (the "lost" soul) and who must return to Him at some point in the future.

Yet there has always existed within Christianity a nonlocal seed that has germinated from time to time in the views of some of the Christian mystics. Not all of them, to be sure, for one cannot say that all mystics, even within the same tradition, conceive of their relation to God in the same way. But many visionaries within Christianity did follow a view that is unequivocally nonlocal.

Some readers may also take exception to the generic way in which the words "mystic" and "mysticism" are used in Part III, which deals with the spiritual implications of the unbounded Universal Mind. Much effort has been expended by modern scholars in creating elaborate typologies of mysticism. We have the nature, soul, and God mysticism of R. C. Zaehner; the subjective and objective mysticism of W. R. Inge; the inward and outward ways of R. Otto; and the extrovert and introvert mysticism of W. T. Stace, to name only a few. Some authorities contend, therefore, that there is no core experience common to all brands of mysticism, and that mystics, like everyone else, see the world through a conditioning filter made up of personal, cultural, and religious elements. Those who take this stand may object to juxtaposing statements of various mystics that seem to espouse a common experience, such as the coming together of God and man described in Part III. Whether or not the statements of mystics from divergent cultural and religious traditions can validly be compared is a complex question. In the end we may do well to remember that, ultimately, *anything* we say about God is a compromised, stepped-down version of reality, including anything said in this book. The mystics know this. They assert with a common voice that *all* descriptions falter, even their own; that God is ultimately incomprehensible; and that He, as Nicholas of Cusa said in the fifteenth century, is "set free from all that can be spoken or thought."

No clearer expression of this point of view in Christianity can be found than in the words of the thirteenth-century mystic, Meister Eckhart, the Father of German mysticism. In his doctrine, the importance of linear time and space is abolished, annulling the barriers between man and God. Eckhart thus arrived at a nonlocal spiritual definition of the self, a joining of man and the Divine. In the Master's words,

> If a soul is to see God it must look at nothing in time; for while the soul is occupied with time or place or any image of the kind, it cannot recognize God. [10]

There are no distant places *to* go in a nonlocal world. Every place is here and every moment is now. Eckhart states:

> I once said here, and it is very true: When a man goes out of himself to find or fetch God, he is wrong. I do not find God outside myself nor conceive him excepting as my own and in me. A man ought not to work for any why, not for God nor for his glory nor for anything at all that is outside him, but only for that which is his being, his very life within him. . . . Some simple folk fondly imagine they are going to see God as it were standing there and they here. Not so. God and I are one in knowing. . . . And the manner of our knowing shall be this, I him as he me, not more or less: just the same. [11]

And,

> Our Lord declares, "Where I am there shall also my servant be," so thoroughly does the soul become the same being that God is, no less, and this is as true as God is God. [12]

There is no great chasm separating man and God that must be bridged by temporal actions. The thing is to *realize* this interrelationship, not bring it into being or "make it happen."

Centuries later another German, the incomparable Rilke, enunciated the same majestic, nonlocal vision of the unity of man and God. "I am the cause that God is God!" he exclaimed. And in an exalted passage Rilke elaborates the sublime theme of indivisibility and oneness which so typified Eckhart's vision:

> What will you do, God, should I die?
> Should your cup break? That cup am I.
> Your drink go bad? That drink am I.
> I am the trade you carry on,
> With me is all your meaning gone. [13]

It is as if Rilke is a later mouthpiece for Eckhart, who expressed the same vision half a millennium earlier:

> God needs me as much as I need him. [14]

This intimate, unitary vision of the relationship of God and man has come to be called the language of *deification*. It is shocking to most Westerners because Western religious thought is overwhelmingly local thought, implictly defining the relationship between God and man as one of enormous separation. By abolishing this distance, Eckhart is moving from a local spiritual tradition that affirms separateness to a nonlocal one that denies it. Yet almost everyone in our culture who has been influenced by the Christian tradition "knows" there is a gulf between man and God, and thus it is not surprising that the language of deification has aroused more enmity from the unmystical than any of the doctrines or practices of the Christian mystics. Indeed, it was no different in Eckhart's day. So controversial were his views that his successors were reluctant to quote him by name. The man himself was officially sanctioned by the Church for these radical statements, and it is likely that the only thing that saved him from being condemned to death was that he died before his trial was completed.

Going beyond the language of deification with its abundant metaphors, many mystics prefer to emphasize the nonlocal nature of God, the Tao, Brahman, Allah, etc., by emphasizing what He (It) is not. The advantage of this *via negativa* is obvious. If something can be specified in detail, we instantly begin to make images of it in our minds. The very instant this takes place, we have made it a local thing, event, or process. But the Almighty is not local, just as we saw the Mind is not local. We cannot "picture" the Mind and we cannot picture God. He is not limited in space nor in time, thus nothing exists outside Him.

Realizing this, mystics of various traditions have attempted to describe the Supreme Being by resorting to negative-sounding descriptions rather than by affirming His positive qualities. Consider, for example, the words of Kabir, the Hindu-Buddhist-Muslim saint: "Some contemplate the formless and others meditate on form, but the wise man knows that Brahma is beyond both." [15] And from the Christian tradition: "God," says the author of *The Cloud of Unknowing,* that anonymously written document which so influenced medieval England, "may well be loved but not thought." [16] Similarly, the daring German mystic Eckhart reveals that the Godhead is "a non-God, a non-Spirit, a non-person, a non-image: a sheer, pure One." [17]

By saying that God is nowhere and that he is no-thing, these visionaries are really stating the opposite: He is everywhere and everything, outside of which nothing can exist. By speaking negatively, they are warning us against making false gods—idols, things we can see and hold, whether in our hands or in our minds in the here-and-now. A god that can be localized in forms is not God. And descriptions that emphasize the negative help keep us in touch with this fact.

For most of us, to speak of thought that takes no familiar form sounds like no thought at all. But in many Eastern and Western traditions ways have been specified to achieve this way of thinking. Within the Christian tradition, Eckhart spoke of *agnosia,* "the supreme level where thinker and thought are one . . . [where] there is no more discrimination . . . [where] all is God." [18] This corresponds to *prajna* in Buddhism and to *ma'rifa* in Sufism—all formless ways of knowing that alone allow the comprehension of a nonlocal reality.

Throughout history the idea that one could have direct, unmediated, and unitary awareness of God has been regarded by the orthodox majority with horror—evidence, as usual, that the nonlocal aspects of reality cannot be apprehended through local approaches which demand, *by their very nature,* that God and man be separate. For example, in Sufism, mystics frequently spoke of union with the Creator. But when al-Hallaj of Baghdad (854–922) spoke as though he were an incarnation of the divine Being, and when he consciously and overtly modeled himself upon Jesus, the full weight of the establishment fell on him: as if accommodating his claim, the religious authorities crucified him.

Another prominent Sufi mystic, Abu Yazid (died 875), was so convinced of his oneness with Allah, which he experienced in *ma'rifa,* that he could exclaim, "Glory to me—how great is my majesty." Abu Yazid was a shrewd spiritual thinker within his tradition of Sufism and influenced its religious character greatly. He propounded the idea of *fana,* the passing away and extinction of the empirical self. This event could be brought about through asceticism and various contemplative techniques. It involved the loss of the dominating awareness of one's own individuality, which is a move from a local to a nonlocal participation in the world.

Few traditions have expressed the nonlocal experience of oneness and unity as profoundly as Taoism. Taoism, however, being a nontheistic tradition, does not speak of God, but the Way, or Tao. The Tao is a Universal Principle that guides the entire world. Taoists do not think of it as most Westerners think of God. Not only is it not a personal entity, it is not even localized in space and time as a "thing" would be. The Tao is inherently nonlocal—beyond space and time, beyond persons, even beyond expression. This radical conception of nonlocal spirituality is epitomized in the following interchange between Chuang Tzu, one of the revered founders of Taoism, and a seeker of the Way, Tung-kuo Tzu:

> Tung-kuo Tzu said, "This thing called the Way—where does it exist?"
>
> Chuang Tzu said, "There's no place it doesn't exist."
>
> "Come," said Tung-kuo, "you must be more specific!"
>
> "It is in the ant."
>
> "As low a thing as that?"
>
> "It is in the panic grass."
>
> "But that's lower still!"
>
> "It is in the tiles and shards."
>
> "How can it be so low?"
>
> "It is in the piss and dung." [19]

God and Man as Cocreator

"In the beginning God created the heavens and the earth." God: the Creator of every*thing,* and in six days. . . . Built into these familiar images is our typically local view: the creation by God, in linear time, of space-occupying things, all within a strict causal framework.

In some spiritual traditions, however, there is no such creator. In Buddhism, for example, we find the doctrine of "dependent origination," according to which everything that exists is constantly changing and depends on everything else. The philosopher of religion, John M. Koller, in his admirable book *Oriental Philosophies,* expresses the implications of this view for the doctrine of creation:

> If whatever is, is dependent upon another, then any kind of "straight line" causality is ruled out. There are no independent beings who are responsible for the existence of dependent beings. For example, the theistic notion that one absolutely independent being—God—created the rest of what exists, and that this created universe depends for its existence upon God, makes no sense in the Buddhist view of dependent origination. Rather, *whatever creates is also created, and the process of creating and being created go on simultaneously without beginning or end.* (emphasis added) [20]

Emerging within the Christian tradition we find similar views. Theologian Carol Ochs, in *Behind the Sex of God: Toward a New Consciousness—Transcending Matriarchy and Patriarchy,* describes a state in which a strictly linear world view, with its rigid cause and effect, is abandoned:

> [If it is true that] there can be no purely one-sided causality . . . this would mean that if God is known through God's effects, then what we become determines in part who God is. God is defined through this world. . . . Moreover, since this world is changing, God too changes, . . . for the act of creation changes and transforms the creator . . . God is both our creator and our creation, our ancestor and descendant. [21]

The Buddhist view of dependent origination and the idea of God and man as cocreators seem to be curiously consistent with certain areas of modern physics. Earlier we saw in the Copenhagen interpretation of quantum physics that the observer is crucial in bringing about certain aspects of physical reality: Before an observation, one cannot speak of an independently existing external world. In their magnificent book, *The Anthropic Cosmological Principle,* physicists John D. Barrow and Frank J. Tipler explore the possibility that the entire universe may owe *all* its properties to the fact that we exist and make observations. How can a mind existing in the present act backward in time to affect creation in the past? Cosmologist John A. Wheeler has described how, in so-called ''delayed choice'' experiments, that this possibility is indeed permitted in princple. According to Barrow and Tipler:

> Wheeler points out that according to the Copenhagen interpretation, we can regard some restricted properties of distant galaxies, which we now see as they were billions of years ago, as brought into the universe now. Perhaps *all* properties—and hence the entire Universe is brought into existence by observations made at some point in time by conscious beings. [22]

In a local world made of separate and independent objects set in passing time, a one-way creation is demanded. In this case the spiritual idea of a unity between the creator and the created is disallowed, and similar ideas from modern physics must be rejected as well. But if one can sense—*deeply* sense—the nonlocal character of the world that God has seemingly built into it, we may find that another view of creation is possible, one in which we are not the end result but active participators in all there is.

In summing up the relation between humankind's spiritual urge and the nonlocal view of reality, perhaps it is useful to begin with the statement of someone outside conventional religious circles. Paracelsus, the Swiss-born alchemist and physician, was a trouble-making, unconventional visionary who stormed through Europe in the sixteenth century, leaving his mark wherever he went. No one ever accused him of being a mystic. Yet he possessed as clear a vision of nonlocal reality as one may find:

> Heaven is man and man is heaven, and all men together are one heaven, and heaven is nothing but one man. You must know this to understand why one place is this way and the other that way, this is new and that is old, and why there are everywhere so many diverse things. [23]

His statement demonstrates that not only may saints and sages of the great religions employ this way of seeing the world, common folk and people far removed from major spiritual traditions also have access to this window on reality.

This is important to keep in mind. The debate over the role of nonlocality in the spiritual life of humanity is *not* a debate between the theistic religions and the monistic creeds. Neither is it a contest between Christianity, Judaism, and Islam against Hinduism and Buddhism. While these faiths differ radically on the importance of a nonlocal view of the spiritual life, the issue is not strictly divisible according to creeds. The questions ultimately resolve into how true the nonlocal way of being is for that person who, uncontaminated by the biases of the great faiths, examines his own place in the world. The test case for nonlocality is not what is contained in canons or holy books, nor even what has been revealed to us by the clearest seers of the various faiths. The truth or falsity of the nonlocal way of being is ultimately decided in the life of the Plain Man or Woman.

But it is increasingly difficult in our world today to find men and women who are ''plain,'' who are unbiased and unprejudiced in their confrontation with the world. Most people on our planet have by now been claimed by one faith or another. Yet we *must* look as far as possible beyond dogma. Can we be Plain Man or Plain Woman? Can we sense the way of the world without filtering it through the religious dogmas that are a ubiquitous part of life today?

It is vitally important to try. Today our world is paradoxically united through the threats of war, terrorism, and nuclear holocaust. We cannot afford any longer to emphasize only our differences. The nonlocal world view is a way of leveling them by means that are consistent not only with good science but with the loftiest mystical insights of humanity—and, we can hope, healing our planet in the process.

Many years ago Aldous Huxley pointed to some disturbing historical facts: In our world's history it is usually the case that the most dreadful religious wars were begun by the time-based religions. Very seldom, by comparison, have the eternity-based religions of the East engaged in crusades, inquisitions, holy wars, witch burning, and the like. These differences are staggering, and they reflect the impulses toward mischief and nastiness that have flowed from the local versus the nonlocal windows on reality.

Not only have the time-based religions engaged in frequent onslaughts against other human beings, they have done the same to our planet by almost uncritically embracing the religion of progress—that modern universal creed which draws its very breath from the belief in the reality of linear time and a rigid causality. The wholesale, unexamined trust in progress is a living-into-the-future. It is a denial of the now. It is a cult that condemns its competitors as guilty of a new heresy: being old-fashioned, backward, and regressive, not chic and up-to-date. Those who do not believe in progress are pejoratively accused of supporting hunger, poverty, illiteracy, and a general state of decline. They are often shouted down as the new blasphemers.

Is there a meeting ground for the time-based, local philosophies and the eternity-based, nonlocal ones? There is, and we have seen it already, as glimpsed at frequent intervals throughout history by the mystics of all the world's great religions. The view of reality of Meister Eckhart and others shows that even within the local, time-based traditions such as Christianity, the nonlocal vision can flourish.

If the religions of humankind are to serve as ethical guides in decreasing the threat we pose to each other and to our planet, it will be because they respond to the nonlocal truths they contain, truths that have eternally been enunciated by the mystics they have produced. Do they insist on being time-based or eternity-based? Do they enshrine spatial separation by installing God "above" and "atop" all else, situating him at some distant point? Or do they abolish this distance by allowing a mutual immanence between Creator and created, between the One and all else? Do they reify the individual, separating the isolated person from God and thus requiring salvation-into-the-future, and do they emphasize the separation of individual persons in a world of "others"? Or do they permit the flowering of the Great Unity?

How our religions respond to these categories—time, space, person—will ultimately determine their usefulness as a viable ethic of the Earth.

10

Beyond Suffering and Death

> *From the tireless labyrinth of dreams I returned as if to my home, to the harsh prison. I blessed its dampness, I blessed its tiger, I blessed the crevice of light, I blessed my old, suffering body, I blessed the darkness and the stone. Then there occurred what I cannot forget nor communicate. There occurred the union with the divinity, with the universe.*
>
> —JORGE LUIS BORGES
> "The God's Script"

Mind and Health

If we recover our soul—if we transcend our limited, isolated self and allow a full flowering of our nonlocal nature, what are the consequences for our health? In search of answers, let us look first at the health experiences of persons in whom this way of being is far advanced.

In 1973 a sixty-year-old Indian named Yogi Satyamurti was buried alive. An earthen pit in the form of a 1.5 meter cube was dug in the lawn on the campus of Rabindranath Tagore Medical College and Hospital in Udaipur, India. Then this sparsely built man entered the pit wearing only a light cotton garment, squatted on the floor, and allowed himself to be sealed in his earthen grave by a roof of bricks and cement mortar. About five liters of water were placed in one corner presumably for drinking, but according to the yogi for the purpose of keeping the air humidified. Then he entered a state of samadhi, or deep meditation, wired to an outside EKG (electrocardiogram) machine that monitored his every heartbeat.

The purpose of this bizarre experiment was to study the changes in heart function that accompany deep meditation. The wires connecting the yogi's body to the machine outside the pit were short enough so that he could not maneuver without disturbing them and causing electrical interference with the EKG tracing that the doctors on the outside could immediately observe.

The yogi remained in the pit for eight days with continuous, uninterrupted monitoring of his heartbeat. The EKG before he entered the pit was entirely normal. Soon after he entered the pit, however, the heart rate began to rise, increasing to 250 beats per minute. This continued until the yogi had been in the pit for twenty-nine hours. Then at 5:15 P.M. on the second day something alarming happened: the electrical activity of the heartbeat suddenly disappeared. Nothing but a straight line could be obtained on the heart tracing. There had been no warning—no prior indication of ischemia, the condition of too little blood flow to the heart, which can usually be seen on an electrocardiogram, and no interference pattern, which would have been the case if he had disconnected or jiggled the wires. Even when increasing the amplification in the EKG machine the doctors could observe no electrical activity whatsoever.

The straight line trace continued until the eighth day. Then, about thirty minutes before the pit was scheduled to be opened, a little electrical disturbance appeared in place of the straight line, finally giving way to a normal appearing cardiac pattern except for a slight increase in the heart rate.

The yogi had predicted the entire scenario. When he went into meditation, he had said, his heartbeat would disappear. After seven days he would begin to come out of deep meditation and his heartbeat would reappear.

When the pit was opened, one of the doctors went immediately to examine him. He was sitting in the same squatting position, in a stuporous condition and very cold. When he was assisted from the earthen pit he began to shiver, which persisted for nearly two hours. He was taken to the laboratory where a followup EKG was entirely normal.

Yogi Satyamurti and his followers were delighted. They felt this amply demonstrated the extraordinary control that yogis have claimed for centuries. As for the doctors, they admitted that "we were left rather perplexed and confused." The EKG changes were not those they had expected. And what about the straight line, indicating no electrical activity of the heart, the sort of EKG one expects from a corpse? The equipment was checked again and again and it functioned flawlessly. Could this intrepid yogi have tampered with the wires? After the experiment other subjects were wired up to check this possibility, but it was not possible for anyone intentionally to disconnect the leads without causing obvious electrical interference patterns on the EKG tracings. "Therefore," the doctors said, "although it is obviously difficult to believe that the Yogi could have completely stopped his heart or decreased its electrical activity below a recordable level, we still had no satisfactory explanation for the EKG tracings before us. . . . The more optimistic amongst us considered this feat to be a marvelous extension of the 'hypometabolic wakeful state of yogic meditation.' . . . The skeptics, however, were inclined to take the whole thing as some cleverly disguised trick." Their report appeared in the *American Heart Journal* in 1973. [1]

What is the connection between buried-alive yogis and our theme of nonlocal mind? It is quite simple. Anyone who has the slightest experience with meditation or with states of profound relaxation knows that in them the world is experienced in a different way, a nonlocal way. That is, in meditation and deep states of relaxation one goes beyond the here-and-now. One takes in space and time not in bits and pieces, but as if one is experiencing infinity and eternity. Many persons describe transcending the sense of the local self or person, merging with the environment or the world at large. These experiences are also accompanied by characteristic changes in the body's physiology, almost always toward a healthier state.

These practices are a shift from the local to a nonlocal way of being. They are "space-time therapies," or therapies of nonlocality. They are a medicine not just for the body, but for the mind as well. For one cannot engage in these practices without bringing into play forces that show, beyond a shadow of a doubt, that the mind and body behave as a unitary whole.

So far, our culture has not fully grasped the important connections between health and the way the mind experiences time, space, and its relation to other persons. We seem to think that these issues are too abstract to matter when it comes to health and illness. Yet nothing could be farther from the truth. When it comes to health, this local view of ourselves as minds limited to the body and the here-and-now can be devastating.

As long as we believe we are time-bound creatures it is impossible for us to rest in the moment. We are always looking backward in time, longing for the unblemished health of our youth, or looking ahead and fretting about what may happen. Though no attention is paid in medicine today to the pathological impact of this distorted way of being in time, the folk wisdom of cultures around the world has known better. Consider, for example, these words of the Chinese poet Po Chü-i from his poem "Resignation," written A.D. 826:

> Keep off your thoughts from things that are past and
> done;
> For thinking of the past wakes regret and pain.
> Keep off your thoughts from thinking what will happen;
> To think of the future fills one with dismay.
> Better by day to sit like a sack in your chair;
> Better by night to lie like a stone in your bed.
> When food comes, then open your mouth;
> When sleep comes, then close your eyes. [2]

Suffering

When we find ourselves suffering it usually isn't long before we hear ourselves crying out, "Why me?"—the eternal lament of the local, hurting self, trapped in the here-and-now. This local view magnifies pain considerably, like pouring salt on a wound. But if *lesser* degrees of suffering can *increase* the sense of the local self, the experience of *severe* suffering and pain can *decrease* it and lead to an expanded, nonlocal sense of being—which can then lead, via a positive therapeutic feedback loop, to a striking *diminution* of suffering.

Many of the innovative therapeutic techniques that have become popular in recent years take advantage of this fact. Biofeedback, hypnosis, meditation, and various forms of relaxation tease one away from the constricted sense of the "I" who occupies this specific moment, this specific place.

Many religious traditions have in different ways emphasized the relationship between the sense of self and time in the quest to understand the nature of human suffering. No one in recent decades has seen these relationships more clearly than Aldous Huxley. In his immortal work *The Perennial Philosophy* he provides the following commentary on the spiritual differences between the local and nonlocal forms of awareness and suffering:

> Where there is perfection and unity, there can be no suffering. The capacity to suffer arises where there is

> imperfection, disunity and separation from an embracing totality. . . . For the individual who achieves unity within his own organism and union with the divine Ground, there is an end of suffering. The goal of creation is the return of all sentient beings out of separateness and that infatuating urge-to-separateness which results in suffering, through unitive knowledge, into the wholeness of eternal Reality. [3]

The relationship between suffering and one's sense of time, space, and person was treated at length in my previous book *Space, Time and Medicine,* which the following passage emphasizes:

> The most hideous aspects of illness are the distortions in spacetime that sick persons experience. These distortions accentuate pain, suffering, and anguish. The spacetime [nonlocal] view of health and disease tells us that a vital part of the goal of every therapist is to help the sick person toward a reordering of his world view. We must help him realize that he is a process in spacetime, not an isolated entity who is fragmented from the world of the healthy and who is adrift in flowing time, moving slowly toward extermination.
>
> To the extent that we accomplish this task we are healers. [4]

Participation in nonlocal being is nothing less than a *spiritual morphine:* it reduces pain and suffering. But unlike a medication, it does not last only until the "injection" wears off, leaving one to face more pain and anguish. The analgesia of nonlocal being is curative, not palliative, for it eliminates the root and the source of suffering: the isolated person, the self, who is trapped in passing time and finite space, drifting toward destruction.

Time, Health, and God

There is no greater obstacle
to union with God
than time.

—MEISTER ECKHART

Does being "more spiritual" make one healthier? A modern authority on the world's religions who was fascinated with this question was Mircea Eliade. From 1933 to 1939 Eliade studied the techniques of yoga in Rishikesh, India. On the basis of extensive observation he came to the conclusion that the rhythmization of the breathing, which is the cornerstone of almost all yogic techniques, might result in striking increases in physical health. Eliade was struck, at Rishikesh and elsewhere in the Himalayas, by the remarkable physical condition of the yogis. They took scarcely any food. One particular yogi he studied was a *naga,* a naked ascetic who spent nearly the whole night in practicing *pranayama,* special breathing techniques, and never ate more than a handful of rice. He had the body of a perfect athlete, Eliade observed, and showed no signs of undernourishment or fatigue. Eliade wondered why he was never hungry. "I live only in the daytime," the *naga* replied. "At night I reduce the number of my respirations by one-tenth." Eliade speculated that if "vital time" could be measured by the number of one's inspirations and expirations, the yogi lived in ten hours only a tenth part of our time, namely one hour, by virtue of the fact that during the night he reduced his breathings to one-tenth of the normal rhythm. "Counted in respiratory hours," Eliade proposed, "a day of twenty-four solar hours only had a length of twelve to thirteen hours for him: thus he ate a handful of rice not every twenty-four hours, but every twelve or thirteen hours.'' Eliade admitted, ''This is only a hypothesis . . . but, as far as I know, there has still been no satisfactory explanation for the surprising youthfulness of the yogis.'' [5]

What are we to make of these sorts of observations? Can yogis really enter another kind of time, one that ''goes slower'' than ours and in which the aging process is retarded? What difference could it possibly make what one *thinks* about time?

A slightly different way of approaching these questions is to turn the situation around and ask what happens *not* when one slows down the experience of time, but when one speeds it up. If there is anything to Eliade's suggestions, we would expect this to have the reverse consequence of making one *less* healthy. Indeed, this seems to be the case. The most dramatic example of this in our current society is in the Type A personality syndrome, wherein time is experienced as if it is running away. There is a pathological sense of fleeting time, never enough of it to go around. The behavior of Type A persons reflects their concern with time in their hurried speech, frenetic gesticulations, and facial expressions. And this heightened time *awareness* is unfortunately translated into an unmistakable ''time *sickness*,'' for Type A persons suffer an increased incidence of death from coronary artery disease, and they die of it at an earlier age than the rest of the population.

That's one side of the time-and-health coin—the exaggerated sense of passing time. But what of the other side? Does *slowing down* one's experience of time *increase* one's health, as Eliade's observations of the yogis suggest?

The most common of all the yogic techniques for achieving spiritual illumination, ''the paradoxical leap outside time'' as Eliade calls it, is the *pranayama,* or the rhythmization of breathing. When one first begins this practice, one is an ordinary creature, a ''profane man—[a] . . . feeble, dispersed slave of [the] body, incapable of a true mental effort. . . .'' Gradually, however, through much effort, this profane man is transformed ''into a glorious Man: possessed of perfect physical health, absolute master of his body and his psychomental life, capable of concentration, conscious of himself." [6]

Now what is important for our consideration here is that, ". . . in working on the respiration, *the yogi works directly on lived time. . . . Pranayama* has been compared to the beatific time of one listening to good music, or the raptures of love, to the serenity or plenitude of prayer. But all these comparisons are inadequate. What is certain is that in progressively decelerating the rhythm of breathing, in prolonging the expiration and inspiration, and increasing the interval between these two elements of respiration, the yogi *experiences a time different from ours.*" [emphasis added]

Through his respiratory rhythm the yogi repeats or simulates "Great Cosmic Time," which contains the periodic creations and destructions of the universe as revealed in Hindu thought. By becoming associated through one's breathing patterns with Cosmic Time, one realizes the relativity and the ultimate unreality of time. But this entrance into Great Cosmic Time does not abolish local time as such for the yogi; for although he has learned that this nonlocal Time is ultimately real, he still lives his earthly life *in* local time. But now, having experienced Great Cosmic Time, there is a difference: Although he is still *in* time, he is freed *from* it.

Here is the paradoxical unification of opposites that it is the genius of Indian thought to see so clearly—two different kinds of time, sacred time and profane time, as Eliade puts it. It is exceedingly difficult for Westerners to believe there could be two kinds of time. We always want to know which is the "right" time, assuming there must be a single kind. But the yogi does not merely *reason* about time, as do we. He *plunges into it* and finds out for himself that time is not a single thing, discovering the realities of sacred and profane time through his own experience. The yogi goes beyond the contradictions of opposites; he rises to the "absolute, intemporal state in which there exists neither day or night, in which there is neither sickness nor old age. . . . [He has become] a *jivan-mukta,* [who] is said to live no longer in time, in our local time, but in an eternal present, in the *nunc stans*." [7]

But the primary goal of the yogic enterprise from the Indian perspective is not perfect physical health but spiritual attainment. Just as the yogi learns that the opposites of sacred time (eternity) and profane time (linear, passing time) come together in unity, he discovers also that the opposites of health and illness come together in the same way. *This* is the ultimate form of health—*not* the acquisition of an unblemished state of physical perfection, but the abolition of the opposites of health and disease, birth and death. A person who understands this stands outside local, passing, profane time, which is the time of illness, suffering, and death. Now he or she is in the graced state of nonlocal, extended time, the time of eternity, the Great Cosmic Time.

One of the most remarkable stories ever to come from the East showing the relationship between health and spirituality concerns the life of one of India's great yogis, Shriman Tapasviji Maharaj. This remarkable man, whose name literally means "Great King, Practitioner of Austerities," was born a warrior-prince in Mogul India in 1770, and when he died in 1955, after living 185 years, he was known and revered all over India. Not only is the length of his life extraordinary, so too are the unusual circumstances under which he chose to live. This incredible man lived much of his life in the practice of self-imposed physical austerities *(tapas)* of almost unbelievable severity. But these in turn were balanced by his profound dedication to the spiritual path, his meditation, and by an ancient technique of rejuvenation called *kaya-kalpa*.

(The combination of physical austerities and meditation is practically unknown in the West, where it is generally viewed as an aberration. But not so in India, where austerities are believed to be a potent means of breaking the hold of the local mind—the ego, the self-contraction, the chronic sense of I which insists that linear, passing time is the only Time there is.)

At the age of 115, Tapasviji Maharaj set out for Darjeeling

on a seven-hundred-mile trip northeast of the forest in which he lived. He walked day after day, mindless of the cold weather, which became severe as he approached the range of hills that lay toward his destination. One day as he stopped to rest near a stream he experienced a desire to perform tapas for three years by standing on only one leg. He chose a spot under a tree that had a low-hanging limb, to which he could reach upward and steady himself. Grasping the limb he stood on his left leg, crossing his right leg over his left knee. He resolved to maintain this position for three years, meditating all the while on his sole source of strength and understanding, Lord Krishna. Closing his eyes, he became absorbed in deep contemplation, firmly planted on one leg and grasping the tree's limb above.

As the days passed, news of his demanding tapas spread to the countryside. Monks and hermits came to gather about him, marveling at the severity of his goal and the dedication he demonstrated. On the sixteenth day he broke his samadhi, his deep meditation, and accepted an offering of a cup of milk that someone had brought. Drinking it, he closed his eyes again and entered the state of blissful solitude. He continued in this routine month after month. Every fifteen days or so he would allow one arm to rest by his side until the next period, but always his left leg remained planted with his right leg crossed at the knee. Following the drinking of a little milk, he once again would enter samadhi.

Throughout all this three-year period thousands of people came to see him and received his blessing, or darshan. No one knew his name, and thus one was given to him: Tapasviji Maharaj.

Something mysterious happened at the end of the three years. One day a stranger happened by, dressed in unconventional garb, and stood quite close to him. The stranger cried out that tapas should not be performed during this present age, as they served no purpose. Tapasviji Maharaj opened his eyes and looked at the stranger, but because he had also taken a vow of silence he said nothing. Then the stranger repeated his words, again drawing no response from the him. The mysterious stranger then caught hold of the right leg of Tapasviji Maharaj and forcefully pulled it to the ground, breaking it at the knee and causing him to cry out in great pain.

"What? Have I caused you pain? I will cure it at once," the stranger said, and he massaged Tapasviji Maharaj's leg. The fractured leg healed and the pain went away. Then the enigmatic stranger announced that he lived in a nearby village, and that Tapasviji Maharaj might come to visit him if he wished. It did not occur to the Saint that this was actually Lord Krishna, the divine object of his love and devotion; thus, after the stranger departed, he immediately resumed his stance as before and reentered his meditation once again. This continued for another few days, at which point Lord Krishna reappeared to Tapasviji Maharaj, this time in a vision. In it he said, "Mahatmaji, I am Shri Krishna. It was I who appeared before you in the guise of a *samnyasi,* and it was I who asked you to terminate your *ugra-tapas*. You must discontinue it now and proceed to Vraja Bhumi."

Now the saintly Tapasviji Maharaj understood that his Lord had graced him, and he ended his tapas at once, walking slowly away from the hallowed spot where Lord Krishna had appeared to him. Vraja Bhumi lay a thousand miles away. Picking up his loincloth and water pot, he began the journey.

Yet even on the journey he continued to practice a form of austerity known as *khade-tapas,* which involved maintaining a standing posture for a long period of time and even sleeping in this position. This he resolved to do until he had reached Mathura, the center of Vraja Bhumi. The specific conditions of this austerity were to be on his feet at all times, to keep his left hand above his head continuously, to remain absorbed in the contemplation of his blessed Lord Krishna at all moments, and to walk leisurely without any inner desire to reach a particular place at a particular time. The practice of these conditions took time: It was not until twenty-four years later that he arrived in Mathura. [8]

The story of Tapasviji Maharaj suggests there is some

formula for living a long, healthy life. If it could work for him, then why not us?

But it is not so simple. The fact is that there are many other saints, perhaps just as holy, who have not fared so well. Many times they seem to have the same aches and pains, vulnerability and devastations, as the rest of us. By any objective assessment of the lives of the saints, good health is decidedly *not* a reward for attaining unity with the Divine, for going beyond the local self, for escaping the confines of the here-and-now. In spite of stories such as we have cited—Yogi Satyamurti who was buried alive for eight days and survived, Eliade's athletic yogi who lived only on a handful of rice and lots of meditation, and Tapasviji Maharaj who lived to be 185 while practicing unthinkable austerities—we can conclude that the achievers of nonlocal awareness and transcendent being are simply *not* supermen and superwomen, by and large. They are *not* hard as steel; they suffer and die from the same irritations that afflict ordinary folk, usually with lives no longer than our own. Even the Dalai Lama, the holy leader of Tibetan Buddhism who is acknowledged as one of the most transcendent beings on earth, wears glasses!

A much more typical example of the relation of physical health to the nonlocal awareness of self is that of one of India's most revered modern mystics, Sri Ramana Maharshi. Maharshi belonged to the tradition of Advaita Vedanta, the ancient Hindu philosophy based on the highest wisdom of the Upanishads. This philosophy points to Reality or Brahman as transcendental and prior to all phenomena. As with other prominent Eastern traditions, it holds that in the highest Realization there is no distinction between the individual, the world, and Brahman or the Divine Reality. Such a self is outside time and beyond the limited, walled-off ego—the nonlocal form of awareness that we have met so many times throughout this book.

Sri Ramana Maharshi's life shows that the attainment of transcendent, nonlocal being does not bring with it a guarantee of perfect health or an exceedingly long life. He died of cancer at his ashram in south India in 1950, and the pain he endured was so great that his moans frequently prevented his devotees from sleeping. During his waking hours Maharshi would treat the situation lightly and joke about it, but the pain in his body was obvious to anyone. And at night he seemed to lose control over himself. In his later days his deteriorating health became a great test both of his devotees' faith in him and of their understanding of the nature of the enlightened adept.

Because his devotees were disturbed by their master's suffering, they asked him to explain what seemed to be a great contradiction: How could a God-realized man such as he, freed eternally from the rounds of birth and death, still feel pain, suffer, and cry out for relief? Did not the enlightened state preclude such misery?

When he heard this question, Maharshi broke into a large smile. With love and compassion he replied: "You take this body for Bhagavan and attribute suffering to him. What a pity! I am not identified with the body. I am the Self. If the hand of the Jnani were cut with a knife, there would be pain, as with anyone else, but because he is not identified with the body, he remains in bliss despite the pain." [9]

Sometimes, then, saints and God-realized adepts suffer like the rest of us. Like us, they may even direct angry outbursts against the Transcendent Reality with which they are so deeply united. While riding through the back roads of Spain during a torrential rainstorm, the bumpy horse-drawn cart of Saint Theresa, one of the great Spanish mystics of the Catholic church, hit a huge pothole and collapsed, throwing her headfirst into the mud. The voice of Jesus then spoke to her from the heavens, saying, "This is how I treat my friends, Theresa." Wiping the mud from her face, she shot back, "No wonder you have so few!" [10]

But there is an all-important difference in the emotional expressions of the God-realized saints and those of unrealized persons, and the difference is contained in the comment of Ramana Maharshi above: There is yet another *level* to their experience of these emotions, such that they are "in them but not of them." They seem to be looking on from outside, neither possessing nor possessed by their emotions. The emotion or the pain does not dominate the realized person. It is experienced, yet it does not control.

One of the great living adepts who has realized the transcendent state of nonlocal being is Da Free John, whose activities and extensive writings are becoming widely known in contemporary American culture. The way in which ordinary human emotion fits into the nonlocal state of being of the God-realized persons can be found in his description of his own emotional responses:

> [Anger] gives me creative force with which to be communicative. . . . If I were incapable of this understanding, this anger, this sensitivity . . . I could not generate a Teaching that is true. . . . All this nonsense about angerlessness, lustlessness, emotionlessness, and lifelessness, is part of the stupidity of exoteric religion. You must be capable of emotion, and you must transcend all self-destructive and other-destructive aspects of emotion through the creative life. [11]

None of which should be interpreted as a justification for the unbridled, wholesale, rough-and-ready expression of any and all emotions. As one can tell from the overall context in which the emotional display of the Great Masters and saints takes place, they are manifesting their emotions with exquisite sensitivity and regard for the effects on others.

One of the most famous examples of how suffering can play a role in spiritual transformation comes from the lore of Buddhism. In the eleventh century the Buddhist adept, Marpa the Translator, brought the Vajrayana Buddhist teachings from India to Tibet. Milarepa, who became Tibet's most beloved saint, became his disciple. In his youth Milarepa had practiced an extreme form of black magic for the purpose of avenging the death of his mother. His manipulation of the weather, it is said, caused the death of more than thirty persons. In despair of his condition he finally traveled throughout Tibet, searching for three years to find Marpa to be his teacher.

At long last he found Marpa harvesting wheat with a scythe on his farm in central Tibet. Throwing himself at Marpa's feet, Milarepa begged him to accept him as his disciple, whereupon Marpa turned on him without a word and beat him nearly to death. Milarepa made no protest against the blows, being prepared to sacrifice his life for the sake of instruction from the famous, revered Marpa. Thus Marpa accepted him as his follower and healed him of his wounds.

But Milarepa's trials were not over. For being a murderer, Marpa subjected him to a great ordeal to cleanse him of his past actions. Marpa required Milarepa to build and dismantle, in succession, nine gigantic stone houses, carrying the rocks from one mountainside to another. But Marpa could not be pleased, and as each house was built he would find fault with it and demand its demolition. As Milarepa toiled with the building, Marpa would hound him unmercifully, demanding that he work faster and sleep less, shouting at him if he displayed the least lapse in energy or attention to detail, even beating him at times. During this entire period Marpa refused to impart any esoteric spiritual teaching to Milarepa, which was the reason he had been sought out in the first place.

But even while inflicting these harsh torments on Milarepa, Marpa displayed deep anguish over his disciple's sufferings. He could be seen weeping secretly in compassion for Milarepa; but he refused to relinquish, out of sympathy for his follower, the strict discipline needed for Milarepa's spiritual purification. In this sense the discipline he meted out was the sure sign of a higher compassion, whose nature could not be judged by a superficial assessment. Marpa later explained the spiritual nature of his motivations in Milarepa's biography:

> I was angered at Milarepa, . . . [but] my anger . . . was not like vulgar, worldly anger. Spiritual anger is a thing apart; and in whatever form it may appear it has the same objective—to stimulate repentance, and thereby to contribute to the spiritual growth of the person. [12]

"Who Hath Sinned?"—Or Why Does Disease Occur?

> *The aim and purpose of human life is the unitive knowledge of God.*
>
> —ALDOUS HUXLEY [13]

As we have seen, there is no automatic association between spiritual attainment and perfect health and longevity. But why not? The question for both the saints and ourselves is: Why does disease occur?

The traditional answer from the Western religious traditions is that suffering is punishment from God for wrongdoings. But when faced with a blind person, Jesus himself denied this association between physical imperfection and sin: "Neither hath this man sinned nor his parents, but that the works of God should be made manifest in him."

Western science has taken another direction. Perhaps the person is blind because of genetic causes, or perhaps because of a congenital infection. In any case, the reason for the blindness is entirely physical.

The East explains disease in yet other ways, one of which is the idea of karma, whereby the negative results of past deeds in previous lives may finally catch up with one. In addition there may be an invisible force at work which Aldous Huxley has called "the final cause, the teleological pull from in front. . . . This teleological pull," Huxley said, "is a pull from the Divine Ground of things acting upon that part of the timeless now, which a finite mind must regard as the future." [14] Against the backdrop of this pull other factors may continue to work—the effects of the genes, the physical environment, or karma, "pushing" from the past into the present. All these influences take place, however, in time—passing time, the time of history. But deliverance—the return to the Divine Ground—"can only be achieved as a consequence of the intervention of eternity in the temporal domain; and eternity cannot intervene unless the individual will make a creative act of self-denial, thus producing, as it were, a vacuum into which eternity can flow." [15]

But, again, answering this call does not automatically mean that the blindness will be cured—and this is the great stumbling block we have been examining. In fact, the blindness may well remain, for it is not the key issue. For when we create the vacuum into which eternity can flow, to continue with Huxley's metaphor, one realizes that "cure" and "longevity" are the concerns not of Eternity but of Time. Perfect physical function is a matter for the flesh, the temporal body. But in Eternity one is not in time, and these concerns cease to be all-important.

The puzzles about why disease occurs are problems only from time-based perspectives—the time in which one is born, endures for a while, sickens, and dies. From the view of eternity, no such problems exist. As Huxley put it, "All shall be well . . . in spite of time, all *is* well, and . . . the problem of evil has its solution in the eternity, which men can, if they so desire, experience, but can never describe." [16]

The Great Question That Sickness Asks

Along with questions such as what is the best medicine or who is the best therapist, illness always asks a larger question: What is the mode of being in which I will take my stand during this illness?—the local mode of the isolated, individual self or the nonlocal mode of the expanded, unitary Person, the state of the One Mind, which is not limited to the here-and-now. In the end, sickness is always a question of being.

The legendary Zen master of the fourteenth century, Bassui, fully understood the need to answer this question of being that is faced by every sick person, and he once responded to queries put to him by a disciple who was sick, in this way:

> You ask me to write you how to practice Zen on your sickbed. Who is he that is sick? Who is he that is practicing Zen? Do you know who you are? One's whole being is Buddha-nature. One's whole being is the Great Way. The substance of this Way is inherently immaculate and transcends all forms. [17]

Bassui was trying to nudge his disciple into a fuller definition of the Self. And elsewhere he wrote to a dying man the following advice, which summarizes the questions always contained in illness, and the best response to them of which we may be capable:

> Your mind-essence is not subject to birth or death. It is neither being nor nothingness. . . . Nor is it something that feels pain or joy. However much you try to know [with your rational mind] that which is now sick, you cannot. Yet if you . . . wish for nothing . . . and only ask yourself, "What is the true substance of the Mind of this one who is now suffering?" . . . you will eventually be freed from your painful bondage to endless change. [18]

What Can We Do to Be Healthy?—Leaving It to Heaven's Care

The "nonlocal therapy" that we have been discussing can be criticized for being too passive. It may sound as if we are advocating nonaction—rolling over dead, doing nothing, letting nature take its course. But these objections are themselves based on a local view of things, suggesting that unless we "do something now," all is lost. But the whole point of the nonlocal view of reality is that there is literally nothing *to* lose and no time to lose it *in*.

But the nonlocal approach to the world is *not* passive. It does *not* preclude taking active measures to prevent illness, nor does it outlaw any particular form of therapy once one is sick. Both in health and in sickness one can act, but one no longer acts out of a sense of desperation or ultimacy. Active efforts may continue, only they are now seen in a different light.

Currently there is great misunderstanding on this point in our society. As a physician to many persons who are dedicated to the idea that "the mind matters" and that spiritual awareness is a major factor in health, I frequently encounter patients who believe, for example, that it is a sign of ethical or moral weakness to have an appendectomy for acute appendicitis or take an antibiotic for pneumonia. They contend that, since the mind and spirit are important in health, one presumably should not fall ill in the first place if one is sufficiently "spiritually aware"; or, if one does become sick, the cure should be attainable through mental or spiritual means. Believing this, they experience a sense of shame and failure on falling ill if they have to resort finally to traditional medical measures.

This extremely idealistic position misses the point. True, all the major spiritual traditions agree that "the higher contains the lower": Our local, physical being—the body—is subsumed *by*, included *in*, and affected *by* our nonlocal, spiritual Self. Yet there is nothing within these traditions that guarantees us a one-to-one correlation between spiritual development and physical health. If we equate the spiritual domain with the physical, we are confusing two different levels of reality. Because these levels differ radically, we are not required to ask the doctor's permission to pray, nor are we obliged to ask God's permission to take an aspirin when we have a headache. Keeping these distinctions clear would go far to allay much of the guilt that is felt when spiritually sensitive persons become sick.

One of the euphemisms that has been popularized in our own culture in recent years is "Going with the flow." There is great wisdom in this bit of advice, I feel. When it comes to health, this phrase implies that there is a natural order of things we would do best to honor at all times, which includes sickness *and* health, mind *and* body, strength *and* weakness. When one allows oneself to really *feel* this way of getting along, then one realizes an inherent rightness in whatever happens.

The ancient Chinese recognized a similar dynamic. One of the best expressions of this eternal approach is embodied in the poem "Thinking of the Past" by the ninth-century Chinese poet Po Chü-i, written in A.D. 833. Po Chü-i, at advanced age, is thinking of former days and reflects on his old friends. They are all dead, in spite of employing the most intrepid techniques of their day to prolong life. In contrast, Po Chü-i never dieted; he yielded to every lust and greed; knew nothing of the popular remedies of his time; and ate and drank with abandon. In spite of this—or because of it—his body is sound and continues to serve him well. He reflects:

In an idle hour I thought of former days
And former friends seemed to be standing in the room.
And then I wondered "Where are they now?"
Only I . . .
Have succeeded in living to a ripe old age.
. . . Already I have opened my seventh set of years;
Yet I eat my fill and sleep quietly.
I drink, while I may, the wine that lies in my cup,
And all else commit to Heaven's care. [19]

Seeing Our Own Death: Gateway to Immortality?

One of the richest sources of evidence of our nonlocal nature can be found in the literature of anthropology, in accounts of persons witnessing their own death by taking outside vantage points. These are not ''near death'' experiences, but experiences of *actual* death. Untold numbers of such stories exist, dating from ancient Egypt to the present day. Anthropologist Gerardus van der Leeuw believes these stories say something important about the nature of human consciousness, its relationship to the body, and the possibility of immortality. ''Man believes in something outside himself,'' van der Leeuw stated, ''which is the pledge of his existence. We may call it consciousness, but then in the sense of *con-scientia,* co-knowledge of our life, objectivization of life from some outside vantage point.'' [20]

A typical story is that of a Winnebago shaman who saw his own deaths and rebirths, recorded by anthropologist K. T. Preuss:

> We were attacked by a company that was on the warpath and killed us all. But I did not know I had been killed. I thought I was running about as usual, until I saw a pile of corpses on the ground, and my own among them. [He goes on living] until I died of old age. My bones fell apart at the joints, my ribs caved in, and I died a second time. I experienced no more pain in death than the first time. This time I was buried in the way then customary. I was wrapped in a blanket and lowered into the grave. There I rotted. I watched the men who were burying me. [21]

The sheer number of these accounts and their historical weightiness led van der Leeuw to observe that,

> This ''outside vantage point'' is always present, even in death. . . . [This suggests that] man as he lives and breathes is assuredly not immortal; nor does he live on after his death. But quite independent of death is the ''it'' which perceives and in perceiving posits an existence beyond death. [22]

But what is this "it," this capacity to observe as an onlooker? The traditional answer from empirical science is that it is an illusion, for the "stuff of consciousness" is simply an expression of what the brain is doing. When the brain dies, consciousness must also evaporate, along with any "it"—the typically local view.

The Western religions have taken a different view. They claim that the "it"—the soul, spirit, mind, or consciousness—is such an intimate part of the body that the flesh *must* die in order for the soul to be liberated. As long as the body survives, the soul cannot really be free to travel homeward to be with the Divine. This idea was classically formulated by Cicero: "The souls of men are divine, and when they depart from the human body, they can return to heaven, the more readily in so far as they have been virtuous and just." [23]

But this does not accord with the anthropological record, says van der Leeuw:

> [This] "it" which sees the man rot, which goes on journeys to heaven and hell, . . . this "it" has *neither body nor soul*. Though it is in no sense an abstraction, it seems to have *nothing to do with the time* that governs our entire life. Our psychophysical ego has a history; it goes through a development. But the perceiver . . . remains *outside the temporal process*: he does not, at our death, vanish from our view (we no longer have any); it is we who vanish from his view. (emphasis added) [24]

Thus the "surviving" onlooker entity, "it," is really not something that survives the death of the body, for it was present before the body came into existence. From this point of view the neurologists are entirely correct: There is indeed nothing that survives the brain, for the "it" was not contained in nor produced by the brain in the first place. This does not deny that "it" does not manifest *through* the brain during life, but that it is not the same thing as the brain, just as an electrical signal might manifest through a television set but is not the same thing as the receiver itself. Destroy the television receiver and the signal remains intact; destroy the brain and the "it" also remains.

This is in contrast to the way we generally think of immortality. In the usual scenario the body dies and "gives off" something—the soul or spirit—via some sort of escape plan previously laid down according to divine sanction. But anthropologically this is not the case: There is simply not a complete union between the body and the "it." The "it" is always prior to and more fundamental than the body. It is always added to the body from the outside and never really bonds with it totally.

Here we have a genuinely nonlocal view of the most essential part of the self: This abiding element—call it what we will—cannot be spatially confined to brain or body, and it is outside of chronological time.

Impressed by the anthropological lore, van der Leeuw believes that "death is . . . indispensable because it admonishes us of the ex-centric ["off-center" or outside] character of our life." [25]

This is why death is the great teacher about immortality. It is no wonder our culture seems so out of touch with immortality, for we are the premier death-deniers—obsessed with "healthy life-styles," "life extension" techniques, or any strategy whatsoever that promises to prolong life. This is one of the great ironies of modern life: that in denying death we deny our own immortality.

There is truth in Archie Bunker's statement, "I'm sorry the guy's dead, but that's life." Death *is* life—because, paradoxically, the only way toward a felt certainty of immortality is through a conscious acceptance of physical death. That is why we must cease all the denial if immortality is ever to become real. We can never believe the pills, potions, and surgery will ever really save us. We will never convince ourselves that technology—such as cryogenic processes, freezing the entire body or some strand of DNA for later use—holds the answer to the problem of death. After all, what happens if the electricity fails?

That is why death must remain, why it *should* remain. There can never be any certainty of immortality without it. As long as death is denied, the experiential fact of immortality can never flower in our lives.

That is one of the ways in which primitive medicine rose above ours. It helped persons go beyond death to immortality because it was not death-denying. ''It is a good day to die,'' the Native American shaman chants. Why good? Because today the fact of immortality will be experienced as never before. And not only today, the day of death, but on all days, because death is not denied on any of them.

No medicine that does not acknowledge the everpresent reality of death deserves to be visited on mankind, whether it is called orthodox, traditional, scientific, allopathic, homeopathic, new-age, alternative, or holistic. Any medicine that denies death is grotesque and cruel because it promotes fear and anguish instead of the ecstatic certainty of immortality. Whether or not death is fully acknowledged as a rightful part of life is the great watershed for the various schools of medicine that ask for our allegiance today. It does not make any ultimate difference whether pills and surgery are traded for herbs, crystals, meditation, homeopathy, chiropractic, massage, or anything else. If death is not an accepted step in therapy, these therapies have no right to claim our attention. Beyond the question of how well they work, the ultimate criterion for all of them is the extent to which they acknowledge and live consciously with death. If they fail this acid test they are unfit to be used on human beings, for they will themselves be death in disguise.

The Maharishi Effect

[We] think of consciousness as being something peculiar to the head, that the head is the organ originating consciousness. It isn't. The head is an organ that inflects consciousness in a certain direction, or to a certain set of purposes. But there is a consciousness here in the body. The whole living world is informed by consciousness.

—JOSEPH CAMPBELL [26]

"[A] discovery that is going to change the world"— that's what the proponents of the Maharishi effect call it. [27] A dreamy, Far Eastern, far-out idea? Hardly. For almost two decades a scientifically disciplined cadre of sociologists, psychologists, mathematicians, biochemists, physiologists, educators, and physicists have patiently studied the effects of what they call "a fourth major state of consciousness [besides sleeping, dreaming, and ordinary wakefulness] that is physiologically and biochemically quite unique." [28] Their approach has been so scientifically rigorous that they have left many of their critics open-mouthed.

What is the Maharishi effect? It is the process wherein the social environment becomes more harmonious, coherent, and transformed as a result of a group of people entering into the experience of "pure consciousness."

And just what is "pure consciousness"? It is a state of mind that apparently has been known for millennia in which the specific contents of the mind—all images and thoughts—take a back seat, leaving only "contentless awareness" as the primary experience. Generally speaking, it is "mind without thought"—

however difficult it may be for our busy, noisy, modern minds to conceive of such a thing.

Many spiritual traditions throughout history have spoken clearly and compellingly about this state of consciousness. For example, it is described as "pure contemplation" by the unknown author of the influential fourteenth century treatise, *The Cloud of Unknowing*. [29] It is called the experience of "pure love" by St. John of the Cross speaking from the sixteenth century. [30] In the Jewish tradition it is called the "Divine Nothingness" or "the Naught" which, according to the founder of the Hasidic movement, the Ba'al Shem Tov, "brings new life to this world." [31] And in the Zen tradition the state of pure consciousness is equivalent to the emptiness of mind that precedes satori, or enlightenment. [32]

What is new in our time, however, is the scientific scrutiny that has been brought to bear on it. What does science have to say about this experience of Divine Nothingness, which can allegedly change the world?

First, this experience is more than an idea or ethereal thought. Important *physical* changes accompany the mindless state. During this state of consciousness, the body becomes very rested, aware, and awake, in spite of a general slowing of its metabolism. Breathing slows drastically, the heart rate drops, skin resistance (a measure of relaxation) increases greatly, and the brain waves as measured on the electroencephalogram (EEG) become quite coherent between different parts of the brain. This is unlike any of the other three common states of consciousness—waking, sleeping, or dreaming. Neither are these changes seen under hypnosis, in which there is no consistent physiological pattern.

But not only does the *body* behave this way consistently during the experience of pure consciousness, the *subjective* experience of this state also follows a typical pattern. As J. T. Farrow reported, in his subjects:

> [There is a gradual change] characterized by an increasingly quiet and orderly state of mind, by an expansion of awareness, and by a reduction of mental boundaries until, all at once, a state of "unboundedness" is reached . . . essentially the same for different people. [33]

During the experience of pure consciousness the sense of the here-and-now fades away and one experiences a merger with something greater than the individual self. This seems to be identical to the state of Nonlocal Reality described earlier, in which the ego and the sense of individual self are transcended.

A turning point in the modern understanding of the state of pure consciousness came in 1974 at a meeting in Arosa, Switzerland, dealing with the scientific research in the area. The meeting was broadly interdisciplinary, with many types of scientists present. The meeting also included Maharishi Mahesh Yogi, the originator of the well-known TM or transcendental meditation technique, which is based on the Vedic tradition of ancient India. During the conference the Maharishi suggested that if only 1 percent of a population began to meditate and experienced pure consciousness, the remaining 99 percent of the population would also be affected. The change, moreover, would not come about by anything the meditators did or said, but only as a result of their experience of pure consciousness in itself. On balance the suggestion was unorthodox, conflicting with the still-current notion that consciousness is entirely inside the brain. But the Maharishi was postulating a nonlocal effect in which consciousness behaved rather like a field, extending beyond the brain and body.

This idea was certainly unorthodox, but it could be tested, and sociologist Garland Landrith did just that. He found that in 1973, the most recent year for which all the needed data were available, among those U.S. cities with populations of 25,000 or more, eleven of them had 1 percent or more of their population practicing the TM technique. As a comparison he selected eleven other cities that had fewer than 1 percent TM participants, but that otherwise resembled the ''1-percent cities'' in terms of population, region of the country, college population, and previous crime rate trends. Landrith found that for the non-1-percent cities the crime rates between 1972 and 1973 followed the national trend, increasing an average of 8.3 percent. But in *each one* of the 1-percent cities there was a *decrease* in the crime rate of 8.2 percent. The likelihood that these findings could have occurred by chance was less than one in a thousand. [34]

Since then Landrith's findings have been confirmed in a score of other studies that have employed a variety of state-of-the-art statistical measures and data-gathering procedures well known to social scientists.

Many critics, however, could not believe these results and suggested that the correlation between TM percentage and deceasing crime rates was coincidental and that some unknown factor or factors were responsible for the results. They argued in effect that ''correlation'' was not enough, and that causality had not been established between TM percentage and improvement in the state of society. The investigators met these objections head-on. Thus in many subsequent studies, extremely sophisticated techniques for determining causality have been employed, such as ''cross-lagged panel analysis.'' These results, taken together, have led Professor David Orme-Johnson to calculate that *the probability of the association between TM percentage and subsequent crime rate decline being a coincidence was . . . less than one in 5 billion.* [35]

At the American Psychological Association Convention in Los Angeles in 1981, psychologist Michael Dillbeck reported results in which he found that among 160 randomly selected U.S. cities, the pattern of associations between TM percentage and crime rate shown in the cross-lagged panel analysis yielded *unambiguous* results: Increases in TM percentage *caused* the crime rate decreases and were not just ''correlated'' with them. [36] Furthermore, Dillbeck later confirmed this result in an additional random sampling of eighty U.S. metropolitan areas. [37]

Since then sociologists have studied the effects of meditation in various cities on a broad variety of indicators of the quality of life. These include the incidence of suicide, homicide, divorce, marriage, death, traffic fatalities, deaths due to alcoholic cirrhosis, rape, robbery, aggravated assault, breaking and entering, larceny, auto theft, unemployment, beer and cigarette sales, total employment, and even environmental indices such as sunny days and carbon monoxide and ozone pollution. [38] Entire states, not just specific cities, have been studied using the same analytical techniques originally used on metropolitan areas. Moreover, experiments have been done in which hundreds of meditators are *inserted* into troubled areas such as foreign countries that are in social and political chaos. [39] Again, employing the most sophisticated analytical techniques available to modern sociologists and criminologists, the results seem to show incontrovertibly that something like Group Mind is at work—that is, when a group of persons intentionally enter a state of consciousness in which the sense of Unbounded Self is experienced, the world, quite simply, changes for the better.

Classical science cannot explain the Maharishi effect. It contains no explanation whatsoever about how something as ethereal as "mind" could affect something concrete and material. But some scientists, looking to new insights and discoveries, have begun to speculate that possible explanations do exist for these phenomena.

There are certain physical systems—lasers are an example—in which a change toward orderliness in a small part can be amplified to change the system as a whole. Scientists call such a change a "phase transition." Other examples of this sort occur widely in nature. Water, for example, exists in a gaseous form as steam. As the temperature of steam is reduced, the water undergoes a phase transition, first to a liquid and then to a solid state, ice, in which its inner organization is extremely orderly. Further examples occur in the phenomena of superfluidity and superconductivity. If the temperature of liquid helium is lowered to a few degrees above absolute zero (-273 K), a remarkable event occurs: The liquid helium undergoes a phase transition into a superfluid, a fourth state of matter that is different from the gaseous, liquid, or solid states. As a superfluid, helium behaves strangely. It can escape through glass containers, becoming "unbounded." Its resistance to flow is zero, and its ability to conduct heat is infinite—properties that are due to the high degree of orderliness or coherence that develops among the helium atoms. All the individual atoms "behave as a single helium atom; the quantum mechanical properties of the atom are allowed to manifest on a macroscopic scale." [40] Other elements such as lead display similar behavior. Reduce lead's temperature to a few degrees above absolute zero, and its electrons arrange themselves in pairs. They become coherent with each other in the sense that their wave nature, fully describable by quantum mechanics, becomes synchronized (more about quantum mechanics and the wave/particle nature of matter later). This new orderliness allows new properties suddenly to manifest, properties such as superconductivity, which did not exist before the coherence occurred.

Could such a thing happen with consciousness states? If a certain number of minds became "more coherent and orderly," could this change society as a whole? Could a subpopulation of individual minds cause a "phase transition" in the collective mind? If consciousness is indeed nonlocal, as we have argued, the problem of spatial separation of minds in individual brains would cease to be a barrier, and there would be an automatic way for information to get around from mind to mind.

This possibility may not be as farfetched as it might seem. Synchronization between the brain wave patterns of subjects has been demonstrated in a number of experiments, even when the subjects are shielded from each other and widely separated. In one experiment, when subjects attempted to communicate by simply becoming aware of each other's presence, they "felt themselves blend into each other" and their brain wave recordings became virtually identical. This suggested to the experimenters that "neuronal fields" can interact and alter each other. [41] And recently mathematical models have been developed suggesting that a network of neurons may exhibit phase transitions to ordered states in the form of persistent firing patterns. [42]

The physicist Lawrence Domash has suggested that the degree of pure consciousness may be related to the degree of the orderliness occurring within the billions of neurons in the brain. The value of TM or similar techniques is that they offer a way to increase the transition from disorderliness to coherence—deexciting the nervous system, or lowering its "mental temperature," as it were—allowing unforeseen effects to emerge from subpopulations to the larger mass. In the case of water, liquid helium, or lead, the "larger mass" would be only the amount of the physical substance in the experimental arrangement; in the case of mind, the "larger mass" would be *all* minds because of the nonlocal nature of consciousness. [43]

Physicist John S. Hagelin, who is recognized as a leader in the attempt to unify the four fundamental forces of nature within a single theoretical framework (a unified field theory), suggests that the experience of pure consciousness—the "unbounded infinite expansion" that persons frequently describe—is actually a direct experience of the unified field. [44] In this state the qualities of wholeness, completeness, and unity of the unified field are directly apprehended by the psyche, and the sense of timelessness and infinity leaps to conscious awareness. These are the moments of the experience of nonlocality, during which, as Whitman said, "It avails not, time nor place—distance avails not."

The implications of the Maharishi effect are obviously enormous. It would be foolish to ignore the possibility that the voluntary entry by a group of persons into a specific state of consciousness can literally change the world for the better. While our politicians and statesmen struggle with the apparently impossible task of achieving world order through traditional means, we must acknowledge that there may be other ways to achieve harmony and order which are disarmingly unadorned and effective. These ways apparently depend on something as simple as sitting down, clearing the mind, and entering a unique state of awareness—no colossal expenditures for defense, no involvement in internecine politics—only the "action of nonaction" and the experience of the "Divine Nothingness," "the Naught," the One Mind, the nonlocal sense of self.

We should not dismiss these ideas because they seem quirky, "Eastern," "psychic," or otherwise offensive to our modern Western minds. Actually, they are generic: They are quintessentially human and are part of the legacy of our species, extending far back into antiquity. If we have lost touch with them, if they offend us, then perhaps *that* is one reason why our world teeters on the edge of an abyss: We have gradually forgotten what it is like to be truly conscious. These ideas invite us once again to awaken, to become fully aware.

Implications for Healing: Era III Medicine

Looking back on the history of medicine since the dawn of the age of science, it is possible to see all forms of therapy settling into certain periods or eras. The first, Era I, could be called "materialistic medicine." In Era I, which has existed for most of the past 100 years in the West, the emphasis is on the material body, which is viewed largely as a complex machine. Era I medicine is guided by the laws of energy and matter laid down by Newton 300 years ago. According to this perspective the universe and all in it—including the body—is a vast clockwork that functions according to deterministic principles. The effects of mind and consciousness are absent, and all forms of therapy must be physical in nature—drugs, surgery, irradiation, and all others.

Era I medicine was a magnificent, awesome step forward in the history of healing. Its accomplishments speak for themselves and are too numerous to name; among them are the "wonder cures" of this century. So significant are these achievements that most persons believe the future of medicine still lies solidly in Era I approaches.

Approximately two decades ago, however, another unique period in the history of healing began to take shape—Era II or "mind-body" medicine. For the first time since the advent of scientific medicine, importance began to be attributed to mind or consciousness in changing the body. It became possible to show scientifically that perceptions, emotions, attitudes, and perceived meanings affected the flesh, sometimes in dramatic ways. These influences were not trivial. Sometimes they were of life-or-death magnitude. All the major diseases of our day—heart disease, cancer, hypertension, and more—were shown to be influenced, at least to some degree, by the mind.

Since the dawn of Era II medicine, a variety of therapies has arisen that depend on "the power of the mind." Biofeedback, meditation, and various forms of relaxation have been shown to cause healthful benefits and are today quite common. They do not conflict with Era I therapies, but exist side by side with them in a complementary relationship. Sometimes Era I and Era II approaches have fused, leading to entirely new disciplines in medicine, such as the developing field of psychoneuroimmunology.

The most important difference between Eras I and II is that the latter attributes a causal power to the mind, which was lacking in Era I. Still the two have much in common: They are both local in nature. This is most obvious in Era I, with its emphasis on a brain-bound mind and an individual body that occupies a specific location in space and a certain span of time. But Era II is also a local approach. It emphasizes the mind and consciousness of the individual person operating on the individual body—all within the local space-time framework of Era I.

Many enthusiasts believe that the mind-body medicine of Era II is the ultimate form of therapy, and that nothing greater will be developed. What kind of medicine could possibly go beyond the action of personal consciousness on one's own body? But, as important an advance as Era II medicine has been, we now can see that it, like Era I medicine, is also limited and incomplete. There are too many phenomena it cannot account for. So today we can envision a more advanced stage—Era III, or "nonlocal medicine."

Era III medicine is the first form of therapy to arise in the age of science that is genuinely nonlocal. Although it, like Era II, emphasizes the causal power of consciousness, it does not regard the mind as operating only within the individual human body or even within a single lifetime of a person. Rather, in the Era III view, minds are spread through space and time; are omnipresent, infinite, and immortal; and are ultimately one, as described throughout this book. And unlike Eras I and II, in Era III health and healing are not just a personal but a collective affair.

Era III medicine constitutes a fundamental shift from all previous periods in medicine because of its nonlocal foundation. This is what makes it so different from what has come before. It cannot be understood in terms of the matter-energy relationships that govern Era I, nor even Era II. In Era III, it is not enough to say that "the mind matters." A further, decisive step is needed: the recognition that the mind not only matters, *but that it is nonlocal*.

It is necessary to postulate an Era III medicine because if one does not, too much scientific data cannot be accounted for, some of which we have already explored. The most challenging examples are nonlocal therapeutic phenomena such as the "prayer study" of cardiologist Randolph Byrd discussed in Chapter 2. Experiments showing the healing action of mind on simple biological systems, such as the Spindrift experiments, are also relevant. And the Maharishi effect shows how entire environments, cities, and countries can be healed through the nonlocal action of mind. These therapeutic effects cannot be explained within a local framework. An additional model of therapeutic reality, one that is nonlocal, is required.

The idea of three separate eras in the evolution of scientific medicine parallels the views of the renowned physicist and cosmologist John. A. Wheeler, who has proposed three eras for physics. In 1984 at the annual meeting of the American Physical Society, Professor Wheeler described Era I physics as characterized by the work of Galileo and Kepler; and Era II by the insights of Newton, Faraday, Maxwell, and Einstein. The latest development, he suggests, is Era III or "meaning physics." Why "meaning physics"? Because when observers "put to use" various measurements at the quantum level of nature, meaning arises, which Wheeler defines as "the joint product of all the evidence that is available to those who communicate." Meaning also demands "the freedom to ask, to choose the question to be put, to decide . . ."—phrases immediately suggesting the importance of mind.

Is there anything in Era III physics that corresponds to Era III medicine? Perhaps. In Era III medicine, the mind exists nonlocally in space and time, extending beyond the individual brain and body. In Era III physics, something similar may be happening. Wheeler suggests that the "community of communicators" who establish meaning "reaches from . . . back in evolutionary history, through today, on into a future that we do not know how to reckon. With an extent vast in time, and conceivably also vast in space . . . the community is enormous that exchanges evidence and builds meaning. . . . What any one communicator does or does not do makes a difference. . . ." Thus the similarities between Wheeler's Era III physics and Era III medicine are striking: His "community of communicators" acting across space and time seem to share the attributes of Universal Mind. [45]

But it is important to recognize that Era III medicine goes beyond physics and cannot completely be derived from it. Even within Era III physics it is still possible for materialists to ascribe mind, consciousness, and meaning to the physical processes within the brain. In this view there is "nothing special" about mind and meaning, because they are only the result of the play of atoms. Era III medicine denies this explanation for mind. It ascribes causal efficacy to the mind, and suggests that minds can do what brains cannot—such as behaving nonlocally in the universe. In Era III, mind, not matter, is primary. As philosopher Willis Harman puts it, consciousness—not matter—was here first.

It is important to recognize that Era III medicine does not discard the majestic achievements of Eras I and II, it honors them and includes them in a complementary, not antagonistic, way. Many people do not understand this. Once they discover the therapeutic potential of the mind, therapy for them becomes "all mind." They may react with horror to Era I therapies, as if it were an ethical or moral compromise to employ anything other than mind in getting well. But just as Era III physics did not eliminate or discard Eras I and II physics, but subsumed them, so, too, does the Era III "medicine of nonlocality" subsume those forms of therapy that went before it. It recognizes that they still apply in certain situations. So Era III medicine does not require us to eliminate the physically based methods of Era I, or the mind-body therapies of Era II; it only adds another dimension that goes beyond their limitations.

Era III Medicine: The Dark Side

Throughout this book we have examined the positive aspects of Era III medicine. Thus we may be tempted to think the effects of mind, acting nonlocally, are always good. But there is a dark side to the effects of nonlocal mind. The medicine of nonlocality is not always "sweetness and light."

The collective, One Mind can bring about disease as well as healing. In it, "the good" has a force, as we have seen, but so too does evil. For example, children reared in abusive families, where a collective negative psychological force seems to hang over the entire family unit, do not grow normally. Not only are they disturbed psychologically, their actual physical development may be dramatically retarded. This process can be extended to larger and larger population units, even to nations. An obvious modern example is Nazi Germany, where a dark force seemed to collectively poison the minds of a great nation. No one is safe from this dark side of the collective mind—evidenced by not only the conscious but the unconscious ways in which almost everyone contributes to the pollution and degradation of the Earth itself, through the heedless consumption of material goods that may be completely unnecessary for one's sustenance.

Psychologist C. G. Jung called this dark side of the human psyche the "shadow." We do not recognize it because it is uncomfortable to confront; it therefore remains buried in the unconscious. The existence of the dark side means there is a paradoxical combination of good *and* evil that exists collectively, manifesting in what Jung called archetypal forces. Because we are unconscious of them and thus do not acknowledge their existence, we are continually at risk of falling prey to them. But Jung continually stressed that we are at risk of succumbing not only to the dark or evil side of ourselves, but to "the good" as well. For when we align ourselves totally with the good, the shadow operates unopposed, and can become the source of unexpected human misery.

One of the weaknesses of the mind-body medicine of Era II is the denial of the shadowy, dark side of ourselves. There is the tendency to refer all that happens to oneself. If my individual mind can influence my own body, if anything negative happens I must have somehow been responsible. As a result of this narcissistic, self-oriented, local perspective, Era II medicine has great difficulty in explaining "why bad things happen to good people."

But in Era III medicine—as Job long ago understood—the situation is different. Because the Universal Mind, in which our individual minds participate, is a *complexio oppositorum*—a complex of opposites, to borrow Jung's term—the light and the shadow exist together as rightful parts of a whole. If we hold to the light and deny the shadow, we will not understand the whole when we experience it, and we will be assailed with the overwhelming force of tragedy, disease, and death. Only through the conscious recognition of the complexity of the whole can we escape being devastated when bad things happen.

And happen they will, in spite of the New Age insistence that "in highest consciousness" we can ultimately experience solely the light and eliminate the shadow, the evil, the bad. Era III medicine denies this possibility in principle, for it recognizes that God and the Universal Mind are a Whole, from which nothing is excluded—even disease and tragedy.

There is, then, a dark and terrible side to the Universal Mind that is seldom acknowledged in California hot-tub philosophies. But to recognize this—to *know* it as deeply as we can—is curiously to be free from the tyranny of the dark side and the periodic buffeting it causes. By becoming *aware* of this unconscious part of ourselves—by "creating consciousness" as Jung put it—we can be free of the labyrinth.

Era III medicine, then, is not utopian. It does not hold out the promise of a troublefree existence. Believers in Era I medicine have occasionally painted rosy scenarios wherein scientific progress will eventually save us; advocates of Era II quite frequently do the same, adding the "power of consciousness" to the power of materialistic science. On the contrary, Era III medicine denies such a vision, claiming that a full participation in the Universal Mind will lead to an awareness of the *Whole*—the light as well as the shadowy sides of human experience. If this prospect seems dismal, it need not be. For, again, the only real prospect of tranquility and happiness lies in the direction of becoming truly conscious—of acknowledging *all* that constitutes the Self, not just part of it.

Nonlocal Mind and "Modern" Medicine

In concluding our examination of the implications for health of nonlocal models of reality, let's take a final look at modern medicine. "Modern" is a word with genuinely local qualities, and "modern medicine" is a local, here-and-now term. It implies a medicine that is recent, up-to-date, contemporary, and "now," as opposed to ancient, antiquated, and old-fashioned forms of therapy whose time has passed. Our concept of "modern" requires a linear time, a time in which the "now" is safely walled off from the past and future—the flowing time of common sense.

But let us recall that in the entire history of science, no experiment has ever been done that shows that time flows. Our fixation on the rigid divisions of time, then, may be illusory. What if the walls separating the past, present, and future are not impenetrable? What if medical actions could exist "for all time" and influence each other, violating the temporal barriers just as the mind has been shown to do? What would the implications for modern medicine be?

Just such a possibility is raised in the nonlocal model of reality proposed by biologist Rupert Sheldrake, whose hypothesis of formative causation we have already examined, in Chapter 8. According to Sheldrake, each new event in nature creates a morphogenetic field, which makes a similar event more likely to occur. The subsequent event, in turn, strengthens the field that gave rise to it, and thus the influence of these fields tends to be cumulative in time. The result is the creation of malleable "habits" in nature, as opposed to unchanging, rigid laws. Events can conceivably be immortal, existing through time in the form of morphogenetic fields, affecting other events and being affected by them.

What if the event in nature were a therapeutic act of some sort—one that "worked"? According to Sheldrake's hypothesis, it would exert a morphogenetic field like any other happening, making similar therapies more likely to "work" in the future. These subsequent events would also add to the strength of the field that shaped them, creating a "therapeutic habit" in nature. We have the architecture here of a great cosmic drug store, a collection of therapies whose power persists in time. Any technique that worked once would be more likely to work again in the future—the ultimate "natural therapy." Over time these fields might become very strong and self-perpetuating, and some therapies might become quite reliable with the passage of time.

If Sheldrake's hypothesis of formative causation proves valid, a certain respect for "ancient methods" might be extremely wise. Rather than casting remedies aside as old-fashioned, we might acknowledge that there may be an inherent value to the fact that they have survived for millennia. "Old-fashioned" would then become a term of respect, not a pejorative one. Whether or not we understand *how* they work would be beside the point; the fact *that* they are still around might say something about their reliability. And it would not be quite right to say that they endure because they work, for if the two-way process of formative causation is correct, the reverse is also true: They work because they endure.

Many ancient therapies come to mind as examples. Various shamanistic methods; mineral baths and waters; herbs; the use of fresh air and sunshine; sleep and dreams; simple exercise and movement; prayer; laying on of hands and similar forms of psychic healing—all these therapies have persisted for thousands of years.

It seems likely that at the dawn of history, when no previous forms of therapy existed on which one could draw, the methods of healing that were first employed were of a minimalist nature—simple activities such as suggestion, talking, touching, looking, or being silent in order to help the sick person. These "tools" involved a certain way of *being*. And, beginning with the first use of these "being therapies," their morphogenetic fields had their infant beginnings and were strengthened with each subsequent success.

But in our time we have virtually disowned the "being" therapies. We refer to them pejoratively as "placebo effects," implying that they are somehow unreal. And whenever possible we have replaced them with "doing" therapies such as drugs or surgery. In so doing we have abandoned many ancient therapies, behind which may stand powerful and invisible morphogenetic fields, which argues strongly for their preservation and respect.

But new therapies arise also, and begin new fields. Sometimes they are so powerful that they persist in time without the guiding, supportive effects of previous fields. The modern anesthetics, vaccines, certain drugs and surgical procedures surely belong to this category. The best methods do not always belong to the past, because genuine novelty and creativity are possible.

But the fields themselves are impartial or neutral. They can be associated with negative, harmful therapies as well as beneficial ones. Trephining, for example—the ancient practice of drilling holes in the skull—persisted for millennia and no doubt was the cause of untold misery and deaths. And the ancient practice of bleeding, which only recently went out of fashion in the West, can be placed in the same category. Even today therapies have a momentum all their own that sometimes exceeds any logic they may contain. A recent example was the use of radical mastectomy in the treatment of breast cancer long after studies showed that a lumpectomy provided equal results. All therapies would be expected to generate morphogenetic fields of their own, whether good or bad. Perhaps that is why so many therapies "die hard."

If we are always swept along to some degree by the momentum of therapies—by the force of their morphogenetic fields, in Sheldrake's terminology—how are we to tell the good therapies from the bad? If the "habits of nature" can overcome rationality and logic, if we can continue to employ therapies in spite of evidence of their uselessness, how are we to guard against self-delusion? It is here that empirical science can be an enormous aid in the form of well-designed clinical trials that test new forms of therapy.

One of the most interesting implications of morphogenetic field theory for medicine is in the domain of thought and behavior. Thoughts, as we have seen, can set up their own morphogenetic fields that help bring events into being. We can imagine what might happen when a form of therapy with a negative effect is used by a powerful healer whose thoughts are highly therapeutic by way of a "positive" morphogenetic field. Here we have negative and positive effects pitted against each other, mediated by their respective fields. It is quite likely that there have been untold numbers of appalling therapies used throughout history that have appeared effective because they were used by healers who possessed intrinsically positive qualities. Thus, flimsy therapies may persist because they are used by good healers. And the situation surely works in the other direction as well: excellent techniques can be sabotaged by the negative morphogenetic field effects generated by appalling "healers."

The Rediscovery of Tradition

The recognition that each therapeutic act generates a morphogenetic field that can gain strength through time suggests that today healers can draw on the abilities of countless healers in the past. This elevates the healing effort to a true collective endeavor with roots in earlier times. The present healer is not alone. His or her efforts summate with those of predecessors through the cumulative effects of morphic resonance and through the two-way process by which morphogenetic fields form. As a consequence, today's healer can summon the power of all past healers to come to his or her aid, as the shaman never failed to do. Healing is thus not the lonely business it frequently seems, for each healer operates within a greater context and so brings the patient back into a large, interconnected community of the spirit.

This should come as a comfort to the patient. Each time a patient has an encounter with a physician, a thousand, a million consultants are instantly summoned to the case. The physician stands shoulder to shoulder with these predecessor-consultants, whose thoughts and actions are present nonlocally—everywhere and every moment—through their morphogenetic fields. If patients were really aware that this healing power could come to bear on their illness, this might generate a powerful expectation of getting well. Such an expectation would generate a healing force that would not be trivial, as it has been shown that expectation can change the body in dramatic ways.

Moreover, the loneliness that every sick person feels could be attenuated by the knowledge that he or she is attended not only by a personal physician but by every other healer in history. The fear and sadness of being stranded in linear time, marooned in a constantly vanishing present that is carrying one inevitably to the brink of extinction, would abate if we could really understand the nonlocal nature of ourselves. For this knowledge would assure us that we can never be alone and isolated, even if we try.

Sheldrake's hypothesis, then, contains implications of a vast, nonlocal, healing network. This nonlocal quality allows us to summon all healers and all healing power, past and present, when we need them.

11

The Revenge of the Good Fairy

But as I see it now, as I feel it, I want my visions to come out of my own juices, by my own effort—the hard, ancient way.

I mistrust visions come by in the easy way. . . . The real insight, the greatest ecstasy does not come from this.

—LAME DEER
Sioux Medicine Man [1]

One's small song—"I like, I dislike," . . . —creates dissonance with the large song. Today we are called to sing a unity song.

—DHYANI YWAHOO
Cherokee Teacher [2]

In the fight between you and the world,
back the world.

—FRANK ZAPPA

Suppose for a moment you could play God and design your own health. You name it, it's yours. A perfect body without the agony of endless aerobics and pumping iron? No more wrinkles or flab? Frozen in time the way you were before you "lost" your figure or your looks? No more early morning aches or pains, no more bunions and hemorrhoids and dandruff, no more of those daily, niggling reminders that, no matter how hard you struggle against it, you are slowly dying? Anything, anything you wish for, is yours for the asking!

It would be like being the central character in a fairy tale. A quite ordinary fellow does something noble and selfless, something far beyond the norm. Suddenly music fills the air, there is a flutter of wings, and the Good Fairy appears. As a reward for his selfless action the man is granted a single wish. Now the Good Fairy has appeared to *you,* allowing you a wish—*any* wish—about your health. No matter what it is, it will be granted.

Remember how many of the Good Fairy stories ended? Frequently they did not have balmy endings where "everyone lived happily ever after." Remember poor King Midas? He wanted everything he touched to turn to gold—and it did. The man's life became a tragedy. Like some omnipotent alchemist, he went around transmuting all the things he loved into the dead, gleaming stuff. No one was safe from him. Or take W. W. Jacobs' tale of the couple who wishes for two hundred dollars and then receives it as compensation for the accidental death of their son. Later they wish him back to life, only to discover that he is so deformed that they hurriedly wish him back to death again. Or the peasants who wish for beardless grain, only to discover that now it can be easily devoured by birds. Then there is the frog who wanted to be as big as an ox, and burst. Even Freud used a story like this in his discussion of wish fulfillment. He told of a quarrelsome couple who was granted three wishes. The greedy wife wishes for sausages, and gets them. This outrages her husband, who wishes them attached to her nose, and it is done. Then they must expend their third and final wish to get things back the way they were.

All these stories end with some ironic twist—the Good Fairy's revenge. They come from anthropologist Mary Catherine Bateson, who has written incisively about the paradoxes and dilemmas contained in unrestrained, wishful thinking. [3] Bateson has discovered that these tales are universal. They can be found in cultures from Native America to China, from Classical Greece to Australia. The fact that they are spread through cultures worldwide suggests that maybe there is a generic wisdom here that we ought to ponder.

If the Good Fairy does appear and offers us any wish whatsoever about our health, these stories suggest we ought to be cautious. Should we wish for an unending, flawless life? Maybe we ought to think it over lest we, too, fall prey to the Good Fairy's revenge. For the Good Fairy stories always embody a deadly trap. They entice us into believing that if a little bit of something is good, more is better—King Midas, for example, asking not for a few gold coins but for an entire golden world. But an instant's reflection—which the central characters in the Good Fairy stories never pause to make—tells us this simply is not true. A little potassium in the blood is a good thing, but too much is deadly. The right amount of hemoglobin is desirable, but too much can cause clogging of blood vessels with strokes, heart attacks, even death. And money? Stories are legendary about the corrupting power of too much of it, in spite of the fact that almost nobody, even the rich, ever thinks he has enough.

No one knows for sure why the Good Fairy stories are so ubiquitous. Bateson suggests that they may play a role in bringing up children. Children are forever wishing for the most outrageous things and never batting an eye. Sometimes they even wish for destructive things, like the death of parents, siblings, or playmates. The Good Fairy stories may have evolved to teach them that there are hidden dangers in wishing unwisely. Children must learn to develop their "reality principle," psychologists state. They must learn to control their urges for omnipotence and their desires for instantaneous gratification. Unless they do they cannot be effective adults and fit harmoniously into society. Perhaps the Good Fairy stories have evolved as a tool to drive these lessons home.

Yet the essential folk wisdom that these stories contain, Bateson states,

> . . . has apparently eluded our culture. . . . After the engineers and technicians have taken some age-old human desire and made it come true, we find ourselves with something that doesn't correspond at all to the remembered vivid longing. The longing for immortality is not satisfied by the possibility of artificially sustained vegetative life or by cryogenics, and even the dream of flight is a far cry from the transport we experience in a sealed container, breathing stale air as we go, after endless delays, from one identical airport to another. The Good Fairy must laugh at our inept wishing. [4]

It is all too easy to find examples that demonstrate this. The wish for the ultimate source of energy—a Good Fairy type of wish if ever there was one—has brought with it Chernobyl and Three Mile Island. But not to worry, the advocates say. We still have two more wishes left, so now we will wish for the perfect human to run the reactors—someone who won't fall asleep on the night shift, someone who knows when to push all the right buttons and turn all the right knobs. What sort of person could possibly function in this way, we never stop to imagine. A robot? Do we then recoil in horror when we see that with our second wish it has been necessary to transform ourselves into automatons in order to get the job done? With our last remaining wish, shall we wish things back to some earlier, more humane level? But now there is the problem with the waste, which won't go away and which nobody wants in their back yard. Oh well, we'll wish it away, too—but no, for the wishes are all used up: the revenge of the Good Fairy.

As it is with the ultimate energy source, so it is with the ultimate offensive weapon. We have that with our first wish. Now, finding ourselves in the shadow of the bomb, which is now in the hands of the Enemy, we are forced to use our second wish on an ultimate defense, such as a "Star Wars" system. This course of events has a mythical quality about it, as Bateson points out. In myths the hero is usually given a magical weapon of some sort, such as a lance or sword. In addition there is a magical defense of some type to go with it—a magic ring, a shield, or a magical phrase to repeat when the going gets rough. And, just like in the myths, today's proponents of the nuclear exchange scenarios believe the hero—ourselves, of course—will come through victorious. We won't even need the third wish, for the defensive system will work as planned. Meanwhile, the Good Fairy is having a belly laugh.

At some deep inner level we may know that the omnipotence that is being exercised in this wild wishing is off base. This inner voice goes by the name of *ambivalence*. Ambivalence works as a balancing device, an antidote to willful excesses. The point is that (1) any dream, taken literally, is likely to be destructive; (2) what we *say* we want is not *really* what we want; and (3) we *know* this at some deep unconscious level. As Bateson says,

> If it is the case that not only is what you want not what you want, but also that, at some level, we already know this, what is the lesson to be learned? One possibility to consider is that, poor forked crea-

> tures that we are, with ambivalence built in to the deepest levels of our personality, we insert the worm into the apple and the weakness into the O-rings ourselves, building the faults into all we do, and this is not avoidable. We have been increasingly committing our lives and the health of the planet to technologies that require perfection, and yet there is good reason to believe that we may sabotage them deliberately in an unconscious reassertion of the truth of fallibility. After all, if only we could learn the lesson of fallibility from the space shuttle disaster, that learning might save our world, and yet the whole machinery of investigation was oriented towards proving that the errors were so specific and exceptional as to make it possible to assign blame and regain the illusion that such errors could not occur in the future. We may have to find ways to reclaim the ancient wisdom intended to protect human beings against fantasies of omnipotence, and to make a virtue of fallibility. [5]

Omnipotence is pathological because it wrenches the wisher out of the natural order of things. It allows him to secede from the world and no longer play by the rules that govern all living things. But when natural patterns are broken, patterns that have evolved over millions of years, we are painfully finding out that it may take more than our one or two remaining wishes to set things right again.

The relevance to health should be obvious. In thousands of ways science and technology have been our Good Fairies. We have been presented with hundreds of wishes for exercising omnipotence. Vaccines, antibiotics, the anesthetics, drugs of all varieties, surgical procedures, diagnostic methods, public health measures—the list is endless. Time after time the music has filled the air and we have heard the flutter of wings as the Good Medical Fairy appears once again with another wish. Unthinkingly we have always snapped them up. If one piece is good, why not the whole cake?

Nature is ingenious at resisting our attempts at omnipotence. As a single example from veterinary medicine, it has recently been shown that the use of chemicals to rid cattle of infections of intestinal parasites backfires in the most surprising way. Not only does the medication work against the cow's worms, when it is excreted in the feces of the cow it continues to do its work. Dung beetles, earthworms, and microbes, which ordinarily decompose the cow's dung into rich fertilizer and which aerate the soil so that it can hold water, no longer do their job. They won't touch the excrement because it is poisonous to them, just as it was poisonous to the parasites inside the cow. The excrement now lies intact in the field, and over time may cover a sizeable portion of the grazing area. Since cows have the habit of refusing to graze to within a certain distance of their excrement, this puts a chunk of pasture out of use and decreases profits for the rancher—which were intended, of course, to be *increased* by the use of the antiworm medication. [6]

All of medicine is alive with similar examples. In a constant war with bacteria, the drug companies produce an endless array of new products. The story is always the same. The drugs are effective initially, but the bacteria win in the end by developing resistance, which calls for the next round of antibiotics. No one, not even the experts, knows where the process will eventually end. The entire program is kept alive by the urge for profits and by the presumption that the pharmaceutical chemists will always have one more trick up their sleeves.

It has been shown that in some of the best hospitals in the country, one-third of all admissions to critical care units is a consequence of iatrogenic diseases—diseases caused by the acts of physicians. And not the acts of *bad* physicians, either, but *excellent* physicians. It is important to realize that we are not witnessing the work of bungling idiots, by and large, when things go wrong in medicine. Even if all the incompetent doctors worked twenty-four hours a day, they could not account for all the iatrogenic problems that deluge us. That is why the system cannot be fixed easily: The trouble cannot be eliminated by eliminating the troublemakers. The widespread belief that blame can be fixed when the system breaks down is a curse today on physicians and patients alike. The malpractice-liability craze is a darkness that hangs heavy over medicine, poisoning our best efforts. It is driven and sustained partly by fact, partly by fiction. It is true that patients *are* hurt, sometimes killed, by the actions of doctors. But when we collectively insist that no fallibility whatsoever should exist in our doctors, our nurses, our hospitals, we have taken a step that is irrational—because sheer perfection is reflected nowhere in the natural order of things. The effect of the denial of fallibility is an awesome escalation of costs, as doctors practice defensive medicine out of fear of being sued. The irony enters when these defensive measures, which are taken to make sure ''everything was done'' and ''nobody gets hurt,'' wind up themselves causing harm.

This is not a defense of malpractice nor a plea to let the quacks go free. Indeed, we ought to *continue* to demand accountability in our health care system, but we should distinguish ''accountability'' from ''perfection.'' The former is found in nature, the latter is not. We desperately need to recognize the difference and give ourselves permission sometimes to be wrong, ending the hypocrisy.

It is true that we *can* redesign the O-rings and improve the dialysis process, as we should. We can also fire the sleepers at the control panels of the nuclear plants, and we can add another hydrogen atom to last year's antibiotic molecule and zap a few more bacteria with the resulting new ''breakthrough'' drug. But in spite of the improvements, we frequently will find that we have only escalated the problem to another level, where future breakdowns will be even more difficult to correct.

If Bateson is right, we are not fooled by our fantasies of omnipotence, for at some deep core we do not believe them. Our inner sense of ambivalence continually tries to balance the situation. At some level we know that fallibility is written into nature as an inherent quality, and that we are not simply "being weak" or manifesting some correctible, characterological flaw.

Bateson asserts that the most poignant Good Fairy tale is that of King Midas, because he is completely cut off from his environment and from those he loves. Ironically, the richer he becomes the more poverty stricken he finds himself, as he is increasingly isolated and estranged. In the Midas tragedy Bateson finds an analogy in our wish for never-changing youthfulness and perfect health. If we were granted this wish we would eventually find ourselves cut off just like King Midas. We would not develop relationships that take time to mature. We would be frozen in time as those we love pass us by, leaving us alone as they move on to form other relationships that are more fulfilling and complex.

As with health, so with sexuality. Unrestrained sexual gratification is an eternal fantasy, but it may also have hidden consequences—in which case, Bateson suggests, AIDS may be another variation on the revenge of the Good Fairy. The same could be said about any other extreme form of indulgence—not only perfect health and unlimited sex, but drug abuse, alcoholism, and gluttony. In fact, it is difficult to think of any form of overindulgence that is *not* associated with physical disease.

Aldous Huxley noted this association almost half a century ago, long before AIDS or "diseases of life-style" were known. He attributed the cause of the suffering that accompanied the unrestrained exercise of pleasure to the local form of being that we have been examining in this book—the ego, the separate I, which dreams of seceding from the natural order and putting its pleasures above everything else. Eventually the revenge of the Good Fairy arrives, as universally predicted, in the form of suffering. Huxley states:

> The urge-to-separateness, or craving for independent and individualized existence, can manifest itself on all the levels of life, from the merely cellular and physiological, through the instinctive, to the fully con-

> scious. . . . Suffering and moral evil have the same source—a craving for the intensification of the separateness which is the primary datum of all creatureliness. . . . Man's capacity to crave more violently than any animal for the intensification of his separateness results . . . in certain characteristically human derangements of the body. [7]

Today it is estimated that the leading causes of death in our society—heart disease, high blood pressure, cancer, and accidents—are directly related to the way we live. Diseases of life-style are predicted to continue to be the biggest menace to life as we enter the next century, evidence that every dream, if fulfilled, is likely to be destructive. The most dramatic expression of the urge-to-separateness at the body's cellular level is cancer. In cancer, certain tissue completely dissociates from the regulatory processes of the remainder of the body. It secedes from the union of other organs and establishes itself in defiant isolation. Like most attempts at pulling out of the natural order, the environment on which it depended is eventually ruined: The body is killed, and the cancer along with it. This unwitting suicide is the price of omnipotence—again, the revenge of the Good Fairy.

The revenge of the Good Fairy can thus be understood in terms of the local and nonlocal ways of experiencing reality that have guided us so far in this book. Ultimately it is a retribution for violating the nonlocal nature of the world. In contrast to the nonlocal mode of being, wherein one senses a connection with the Whole that transcends local time and space, omnipotence is an expression of a local mind denying its unitary, interdependent nature. It is the craving of the single, isolated individual who values passing time and whatever gratification it may contain above all else. The omnipotent, power-hungry person is divorced from the context of eternity and the natural order that includes all persons and things, all places and moments.

The solutions to the problems that confront us will come

only when we learn to move away from the local to the nonlocal view of reality. If minds *are* nonlocal, as this book contends—if individual minds communicate, if they are fundamentally part of a larger Mind that is boundless in space and time—then there is hope that a few awakening minds can make a difference in the larger Mind of humanity. In which case the fantasies can be given up and we can regain our collective sanity. And if minds communicate we must be careful what we dream, what we fantasize, what we wish for.

Then and only then can we escape the revenge of the Good Fairy.

Epilogue

There are, then, two ways of perceiving our place in the universe. One way, our familiar picture, is soulless. It tells us we are limited creatures whose minds are confined to our brains and bodies and to the present moment. When all our moments run out, our bodies will perish and carry us into the abyss of nothingness.

The alternative to this local, grotesque, and soulless view is the nonlocal picture we have emphasized throughout this book. In it, soul-like characteristics emerge—qualities that defy confinement to individual brains and bodies, and which appear to be unbounded in space and time. As we have seen, immortality and the merger of single minds into a Universal Mind seem to be the consequences of this view. As a further consequence, *if minds cannot be bounded in space and time, we must be prepared to admit—whether we feel squeamish about it or not—that we are endowed with the godlike characteristics of immortality, omnipresence, and unity*.

But, on the whole, we are not yet fully capable of acknowledging the god-qualities we contain—our ''omniconsciousness'' as Jung called it. To compound this blindness, too often our religious traditions have driven home the message that we are despicable, unworthy creatures with no redeemable qualities of our own. And when, throughout history, individuals have arisen who have clearly sensed their inner divinity, they have frequently been accused of heresy and blasphemy and have been treated inhumanely or killed, particularly in the West. These woeful facts reveal the unbelievable lengths to which we will go to hide from our true nature, and they show that we have habitually confused *participation in* the Godhead with *usurpation of* the Divine.

Thus was Jung led to the conclusion, late in life, that "the Christian nations have come to a sorry pass; their Christianity slumbers and has neglected to develop its myth further in the course of the centuries. . . . People do not realize that a myth is dead if it no longer lives and grows. . . . Our myth has become mute, and gives no answers. The fault lies not in it as it is set down in the Scriptures, but solely in us, who have not developed it further, who, rather, have suppressed any such attempts." [1]

Today the question is how to reawaken the old myths, or how to devise new ones. It seems at least theoretically possible to stir the old myths to life—to take seriously the Old Testament idea of the "divine marriage" and the subsequent notion of "Christ within us"; to take Jesus at His word when He said in John 10:34, "Is it not written in your law, 'I said, you are gods'?" Yet there is such enormous inertia and resistance to revitalization. And while we wait for the rekindling of our myths to occur, we border on global catastrophe for the lack of a world-sustaining vision.

Today the resistance to reawakening to our inner divinity, of recovering our soul, comes not only from religion but from science as well—and here the Western religions and science have paradoxically become unwitting bedfellows. Both assure us we are frail, weak creatures who are born to suffer, decay, and die—the familiar local scenario. Both hold out the promise of salvation: one offering it in the form of God's redeeming generosity, the other in the form of scientific progress. But both have stripped us of our omniconsciousness and our soul, becoming dark allies in this morbid process.

There is a story of a student who came to a rabbi and said,

''There were men in the olden days who saw the face of God. Why don't they see it any more?'' The rabbi responded, ''Because nowadays no one can stoop so low.'' [2] What paradox, that to see the highest we must be willing to bend, perhaps to fall completely! Today we are reminded of the need to stoop low by one gigantic, ineluctable fact: The Earth does not need us and might, in fact, be better off without us. We are the greatest threats to its survival—its air, soil, and water—and it requires no great imagination to see the end of our species in sight. If our ravaging of our planet continues, we will be brought unimaginably low—without sustainable water, air, or food. Could it turn out that in this involuntary stooping, of being brought low, there is contained a chance to see, like those of olden times, the face of God? Could the impending ecological disasters hold the opportunity and impetus to revitalize our old myths or perhaps create new ones? Are we being taught the lesson we most need to learn—that we *have* a soul, that we *are* onmniconscious—by the Earth itself? *Are we being offered a final chance of recovering our own souls?*

At this very difficult moment there is a great temptation to resort to the comforts of the old religions—of resigning ourselves to Armageddon or fixing our gaze not on the Earth and our problems but on some distant heaven; or, on the contrary, planting our faith in dreamy euphemisms such as ''science will save us.'' Neither choice is acceptable. A grander option exists: the awakening to a new kind of consciousness that goes beyond the person and, even, beyond the Earth itself. As the great Jesuit mystic and scholar, Pierre Teilhard de Chardin, put it,

> For men upon earth, all the earth, to learn to love one another, it is not enough that they should know themselves to be members of one and the same *thing;* in ''planetising'' themselves they must acquire the consciousness, without losing themselves, of becoming one and the same *person*. [3]

But how? Importantly, if we are to "planetise" ourselves we must temper the hubris and the unmitigated tenacity with which modern science has chosen to deny anything "higher" in creation, including ourselves. We must acknowledge that one can still do science and "acquire the consciousness" by which we can recover our own soul and that of our planet, without losing ourselves in dreamy mysticism. This is *not* an impossibility: Many of our greatest scientists stand as examples, and we have examined their views. Many others could be mentioned, such as Eddington, Planck, Heisenberg, Jeans, Pauli, and de Broglie, whose views on the relationship between science and spirituality have been examined by psychologist Ken Wilber in his important book, *Quantum Questions: Mystical Writings of the World's Great Physicists*. [4]

Yet we must be cautious not to homogenize science and religion, for nothing would be more disastrous. Rather, we must insist that science and religion stand side by side, respecting the domain of the other, and give up the incessant battle to usurp each other's territory. When this usurpation happens—as when theologians pontificate about the fine points of evolutionary theory, the age of the Earth, or the accuracy of carbon dating techniques, or when scientists authoritatively assert that "God is dead" or that there is no intrinsic meaning in the universe, as if they have just reached these conclusions from their pointer readings—both religion and science are diminished. The *differences* between science and religion are just as important as any convergences that may exist, and must be maintained. [5]

This book has tried to demonstrate, by focusing on the behavior of the psyche, how these two domains might get on with each other in a way that preserves the highest aims of each. As a single example, we have seen that science may safely say, based on much data, "There seem to be qualities of the human psyche that demonstrate nonlocality in time and space, and which appear unbounded." Religion at this point may respond, "Of course, we knew it all along! Our term for these 'qualities' is the 'soul,' "—each side saying only what it is justified in stating. Although this dialogue seems simple, it has never worked, for in practice both scientists and religionists have been unable to resist the temptation to try to demolish the other with unimpeachable arguments. As a consequence of this pettiness, which has endured for 300 years, not only has the holy become hollow, most persons by now harbor a distrust of science, a suspicion based in part on its pretense of having demonstrated the shallowness and unnecessariness of religious thought. Which side has been more foolish is difficult to say. In any case, the world can no longer afford such childish disputes; for if we persist in these arguments, eventually there may be no one left to argue.

The great Indologist, Heinrich Zimmer, said:

> [The] divine being in us can never . . . be asleep or away on a voyage. . . . The rationalistic unbeliever, for whom all this does not exist, has no gods left in his heart; that is his fall from grace. Instead, he is tormented by the devils of his brain: "Chance" or "force of circumstances" or "ineluctable causality" . . . for he must seek outside himself all the demons which he does not wish to find and resolve in himself. "Where there are no gods," Novalis said, "there are ghosts." [6]

This book, then, is a plea for a recovery of the soul and a return to the Divine within. If it was once feared that one had to make a choice between intellect and emotion in answering the call of the Divine, then surely we may lay this fear to rest. If we choose to hearken exclusively to the soulless messages of science that have predominated up to now, for fear of compromising our reason, then this choice can only be described as self-indulgence—for within science today we see the unmistakable emergence of a new latitude for the human spirit which simply did not exist in the recent past. As the eminent physicist Werner Heisenberg was led to say, " 'consciousness' and 'spirit' can be related in a new way to the scientific conception of our time." [7]

In the end we can choose to continue to believe that we are local, isolated, doomed creatures confined to time and the body and set apart from all other human beings. Or we may elect to open our eyes to our immortal, omnipresent nature and the One Mind of which we each are a part. If we choose the former, nothing will save us. If, however, we choose to awaken to our divine Self, we face a new dawn.

Notes

Epigraph, p. xiii: Joseph Campbell, *The Inner Reaches of Outer Space: Metaphor as Myth and as Religion* (New York: Alfred van der Marck Editions, 1986).

Preface

1. Henry Margenau, *The Miracle of Existence* (Woodbridge, CT: Ox Bow Press, 1984), p. 72.
2. Ken Wilber, *Spectrum of Consciousness* (Wheaton, IL: Quest, 1979).
3. Paul Davies, *God and the New Physics* (New York: Simon & Schuster, 1983), p. 80.
4. Ibid., p. 133.
5. John A. Wheeler, ''Beyond the Black Hole,'' *Some Strangeness in the Proportion*, H. Woolf, ed. (Reading, MA: Addison-Wesley, 1980); cited in Paul Davies, op. cit., pp. 39, 111.
6. John C. Eccles, *The Human Psyche* (New York: Springer International, 1980), p. 25.

Chapter 1, The Reach of the Mind

1. Richard Erdoes, *Lame Deer—Seeker of Visions* (New York: Simon & Schuster, 1972), p. 155.
2. Candace B. Pert, quoted in David Kline, ''The Power of the Placebo,'' *Hippocrates,* May/June 1988, p. 26.
3. Candace B. Pert, ''The Wisdom of the Receptors: Neuropeptides,

the Emotions, and Bodymind,'' *Advances* 3:3, (1986), pp. 8–16.

4. G. Richard Smith et al., ''Psychological Modulation of the Human Immune Response to Varicella Zoster,'' *Archives of Internal Medicine* 145 (1985), pp. 2110–2112.

5. S. Black, J. H. Humphrey, and J. S. F. Niven, ''Inhibition of Mantoux Reaction by Direct Suggestion Under Hypnosis,'' *British Medical Journal* 1:5346 (June 1963), pp. 1649–1652.

6. G. Richard Smith and S. M. McDaniel, ''Psychologically Mediated Effect on the Delayed Hypersensitivity Reaction to Tuberculin in Humans,'' *Psychosomatic Medicine* 46 (1983), pp. 65–70.

7. G. R. Smith et al., ''Psychological Modulation . . .,'' p. 2111.

8. Paul Davies, *God and the New Physics* (New York: Simon & Schuster, 1983), p. 86.

9. Lawrence LeShan, *From Newton to ESP* (Wellingborough, Northamptonshire: Turnstone Press, 1984), pp. 100–101.

10. John Beloff, ''J. B. Rhine on the Nature of *psi*,'' *J. B. Rhine: On the Frontiers of Science,* R. K. Rao, ed. (Jefferson, NC: McFarland, 1982), pp. 97–110.

11. J.B. Priestley, *Man and Time* (London: W. H. Allen, 1978), p. 245.

12. Reported in *Brain/Mind Bulletin* 12:7 (March 1987), p. 1; and Andrew Greeley, ''The Impossible: It's Happening,'' *Noetic Sciences Review*, Spring 1987, pp. 7–9.

13. S. Weisburd, ''The Spark: Personal Testimonies of Creativity,'' *Science News* 132, Nov. 7, 1987, p. 298.

14. Arthur Koestler, *The Act of Creation* (New York: Dell, 1964), p. 177.

15. Jacques Hadamard, *The Psychology of Invention in the Mathematical Field* (Princeton: Princeton University Press, 1949), pp. 142–143, quoted in A. Koestler, op. cit., p. 171.

16. J. Hadamard, op. cit., p. 85, quoted in A. Koestler, op. cit., p. 172.

17. J. Kendall, *Michael Faraday* (London: Faber, 1955), p. 138, quoted in A. Koestler, op. cit., p. 170.

18. A. Koestler, op. cit., p. 208.

19. Cited in John Chesterman, *An Index of Possibilities: Energy and Power* (New York: Pantheon Books, 1974), p. 186.

20. Max Knoll, ''Transformations of Science in Our Age,''

Man and Time, Joseph Campbell, ed., Bollingen Series XXX:3 (Princeton: Princeton University Press, 1957), p. 270.

21. William Irwin Thompson, *Evil and World Order* (New York: Harper and Row, 1976), p. 81.

22. In Jamake Highwater, *The Primal Mind: Vision and Reality in Indian America* (New York: New American Library, 1981), p. 96.

23. Bernard d'Espagnat, *In Search of Reality* (New York: Springer-Verlag, 1983), p. 102.

24. Jeremy Rifkin, *Time Wars* (New York: Henry Holt, 1987).

25. Carl G. Jung, *Memories, Dreams, Reflections* (New York: Vintage, 1968), p. 302.

26. Carl G. Jung, *Psychology and the East,* R. F. C. Hull, trans. (Princeton: Princeton University Press, 1978), p. 69.

27. Ibid., p. 131.

28. Ibid., p. 132.

29. Ibid., pp. 126–127.

30. Aniela Jaffé, ed., *C. G. Jung: Word and Image;* Bollingen Series XCVII: 2 (Princeton: Princeton University Press, 1979), p. 214.

31. C. G. Jung, *Memories, Dreams, Reflections*, p. 326.

32. C. G. Jung, *Psychology and the East,* R. F. C. Hull, trans. (Princeton: Princeton University Press, 1978), p. 63.

33. Ibid., p. 63.

Chapter 2, The Power of Nonlocal Mind

1. John G. Neihardt, *Black Elk Speaks* (New York: Washington Square Press, 1972), p. 36.

2. Howard Wolinsky, "Prayers Do Aid Sick, Study Finds," *Chicago Sun-Times,* January 26, 1986, p. 30.

3. "Cardiologist Studies Effect of Prayer on Patients," *Brain/Mind Bulletin* II:7 (March 1986), p. 1; see also, Randolph C. Byrd, "Positive Therapeutic Effects of Intercessory Prayer in a Coronary Care Unit Population," *Southern Medical Journal* 81:7 (July 1988), pp. 826–829.

4. H. Wolinsky, op. cit., p. 30.

5. Aldous Huxley, *The Perennial Philosophy* (New York: Harper Colophon Books, 1945), p. 2.

6. Ibid., p. 5.

7. Ibid., p. 7.

8. Ibid., p. 9.
9. Ibid., p. 10.
10. Ibid., p. 11.
11. Ibid., p. 12.
12. Ibid., p. 14.
13. Ibid., pp. 1–20.
14. Ibid., pp. 20–21.
15. Ibid., p. 20.
16. Robert Owen, *Qualitative Research: The Early Years* (Salem, OR: Grayhaven Books, 1988), p. 22.
17. Ibid., p. 76.
18. Ibid., pp. 22–23.
19. Ibid., p. 85.
20. Ibid., p. 89.
21. Albert Einstein, *Ideas and Opinions* (New York: Crown, 1954), pp. 38–45.
22. Lawrence LeShan, *The Medium, the Mystic, and the Physicist* (New York: Viking, 1974), p. 111.
23. Jeanne Achterberg, *Imagery in Healing* (Boston: New Science Library, 1985).
24. L. LeShan, op cit.
25. W. T. Stace, *Mysticism and Philosophy* (New York: J. B. Lippincott, 1960), p. 66.
26. Rudolf Otto, *Mysticism East and West* (New York: Meridian Books, 1957), p. 67.
27. Kurt Goldstein, "Concerning the Concept of Primitivity," *Primitive Views of the World*, S. Diamond, ed. (New York: Columbia University Press, 1964), p. 8; quoted in L. LeShan, op. cit., p. 59.
28. L. LeShan, op. cit., pp. 65–66.
29. Ibid., pp. 106–107.
30. Ibid., p. 107.
31. Bernard S. Siegel, *Love, Medicine, and Miracles* (New York: Harper and Row, 1986).
32. A. Huxley, *The Perennial Philosophy*, p. 57.
33. L. LeShan, op. cit., p. 113.
34. Ibid., p. 114.
35. Brendan O'Regan, "Healing, Remission and Miracle Cures," *Institute of Noetic Sciences Special Report*, May 1987, pp. 3–14.
36. Ibid., p. 11.

37. Ibid., p. 9.

38. Ibid., p. 9.

39. Larry Dossey, *Medicine and Meaning,* in progress.

40. Jonas Salk, quoted in Brendan O'Regan, "Healing: Synergies of Mind/Body/Spirit," *Institute of Noetic Sciences Newsletter* 14:1 (Spring 1986), p. 9.

41. Larry Dossey, "Holistic Health: A Critique," *Beyond Illness* (Boston: New Science Library, 1984), pp. 163–191.

Chapter 3, Bodymind

1. For a thorough discussion of the historical attitudes surrounding syphilis and the way scientific facts are generated, see the little-known, excellent work by Ludwik Fleck, *Genesis and Development of a Scientific Fact,* T. J. Trenn and R. K. Merton, eds., F. Bradley and T. J. Trenn, trans. (Chicago: University of Chicago Press, 1979). Fleck contends that scientific facts are not discovered but invented.

2. Ibid., pp. 1–18.

3. M. Vera Bührmann, *Living in Two Worlds: Communication Between a White Healer and Her Black Counterparts* (Capetown: Human and Rousseau, 1984), p. 15.

4. Candace B. Pert, "The Wisdom of the Receptors: Neuropeptides, the Emotions, and Bodymind," *Advances* 3:3 (1986), pp. 8–16.

5. D. T. Krieger and J. B. Martin, "Brain Peptides," *New England Journal of Medicine* 304:15 (April 9, 1981), pp. 876–885.

6. C. B. Pert, op. cit., p. 8ff.

7. Judith Hopper and Dick Teresi, *The 3-Pound Universe* (New York: Dell, 1986), p. 386.

8. Carl G. Jung, quoted in *Artifex* (publication of Archaeus Project, Minneapolis, MN) 5:6 (December 1986), p. 19.

9. Paul Watzlawick, ed., *The Invented Reality* (New York: W. W. Norton, 1984).

10. Robert Rosenthal and L. Jacobson, *Pygmalion in the Classroom: Teacher Expectation and Pupils' Intellectual Development* (New York: Holt, Rinehart, and Winston, 1968). See also R. Rosenthal and L. Jacobson, "Teacher Expectations for the Disadvantaged," *Scientific American* 218:4 (April 1968), pp. 19–23. For a thorough review of this area of experimental psychology, see R. Rosenthal and

D. Rubin, "Interpersonal Expectancy Effects: The First 345 Studies," *The Behavioral and Brain Sciences* 3 (1978), pp. 377–415.

11. Robert Rosenthal, *Experimenter Effects in Behavioral Research* (New York: Appleton-Century-Crofts, 1966).

12. L. Cordaro and J. R. Ison, "Observer Bias in Classical Conditioning of the Planaria," *Psychological Reports* 13 (1963), pp. 787–789.

13. P. Watzlawick, op. cit., p. 104.

14. Ibid., pp. 104–105.

Chapter 4, Creatures Great and Small

1. Richard Erdoes, *Lame Deer—Seeker of Visions* (New York: Simon & Schuster, 1972), p. 119.

2. John G. Neihardt, *Black Elk Speaks* (New York: Washington Square Press, 1972), p. 4.

3. Teri C. McLuhan, *Touch the Earth* (New York: Simon & Schuster, 1972), p. 23.

4. Mircea Eliade, *Shamanism* (Princeton: Princeton University Press, 1964), p. 459.

5. Ibid., p. 460.

6. Ibid., p. 385.

7. Ibid., p. 94.

8. Ibid., p. 95.

9. Ibid., p. 95.

10. S. F. Nadel in Eliade, op. cit., p. 31.

11. Ibid., pp. 29–31.

12. Ibid., p. 29.

13. Ibid., p. 29.

14. Lyall Watson, *Gifts of Unknown Things* (New York: Simon & Schuster, 1986).

15. J. Allen Boone, *Kinship with All Life* (New York: Harper and Brothers, 1952), quoted in Bill Schul, *The Psychic Power of Animals* (New York: Fawcett, 1977), p. 5.

16. Quoted in Schul, op. cit. pp. 5–6.

17. Ibid., p. 6.

18. Ibid., p. 161.

19. Ibid., p. 161.

20. Ibid., p. 48.

21. Ibid., p. 47.

22. Ibid., p. 49.

23. Ibid., p. 49.

24. Bill Schul, *The Psychic Power of Animals* (New York: Fawcett, 1977).

25. Joseph Banks Rhine and Sara R. Feather, "The Study of Cases of 'psi-trailing' in Animals," *The Journal of Parapsychology* 26:1 (March 1962), pp. 1–21.

26. Schul, op. cit., p. 52.

27. Roy Bedichek, *Karánkaway Country* (Austin: University of Texas Press, 1974), p. 39–50.

28. Schul, op. cit., p. 56.

29. E. Friedmann et al., "Animal Companions and One-year Survival of Patients after Discharge from a Coronary Care Unit," *Public Health Rep.* 95 (1980), pp. 307–312. See also A. M. Beck and A.H. Katcher, "A New Look at Pet-facilitated Therapy," *Journal of the American Veterinary Medical Association*, 184 (1984), pp. 414–421; and P. Gunby, "Patient Progressing Well? He May Have a Pet," *Journal of the American Medical Association (News)* 241 (1979), p. 438.

30. Gustav Eckstein, *Everyday Miracle* (New York: Harper and Brothers, 1940).

31. In M. Hornig-Rohan and S. E. Locke, *Psychological and Behavioral Treatments for Disorders of the Heart and Blood Vessels* (New York: Institute for the Advancement of Health, 1985), p. 176.

32. T. C. McLuhan, *Touch the Earth*, p. 6.

33. Lynn White, Jr., "The Historical Roots of Our Ecological Crisis," *The Art of Reading*, E. Gould, R. DiYanni, and W. Smith, eds. (New York: Random House, 1987), pp. 683–691.

34. Ibid., p. 691.

35. Ibid., p. 690–691.

36. Lyall Watson, "Natural Harmony: The Biology of Being Appropriate," lecture delivered to The Isthmus Institute, Dallas, TX, April 1989.

Chapter 5, The Immortal, One Mind

1. Arthur Eddington, "Defense of Mysticism," *Quantum Questions: Mystical Writings of the World's Great Physicists*, Ken Wilber, ed. (Boston: New Science Library, 1984), p. 206.

2. James Jeans, *Physics and Philosophy* (New York: Dover, 1981), p. 204.

3. Erwin Schrödinger, *What Is Life? and Mind and Matter* (London: Cambridge University Press, 1969), p. 145.

4. Erwin Schrödinger, *My View of the World* (Woodbridge, CT: Ox Bow Press, 1983), pp. 31–34 (original English translation Cambridge: Cambridge University Press, 1964).

5. E. Schrödinger, *What Is Life?*

6. Ibid., p. 1.

7. Ibid., p. 133.

8. John A. Wheeler, in Anthony P. French and P. J. Kennedy, *Niels Bohr: A Centenary Volume* (Cambridge: Harvard University Press, 1985), p. 329.

9. John A. Wheeler and J. Mehra, eds., *The Physicist's Conception of Nature* (Boston: D. Reidel, 1973), p. 244.

10. Jacob Bronowski, *The Common Sense of Science* (Cambridge: Harvard University Press, 1955), p. 77.

11. E. Schrödinger, *What Is Life?*, p. 137.

12. *Mundaka Upanishad,* I.1.6, John M. Koller, trans., in J. M. Koller, *Oriental Philosophy* (New York: Charles Scribner's Sons, 1985), p. 28.

13. *Chandogya Upanishad,* VI.9.4, *The Thirteen Principal Upanishads,* Robert E. Hume, trans. (New York: Oxford University Press, 1921, repr. 1975), p. 246 in reprint.

14. *Chandogya Upanishad*, VII.7.1, *The Principal Upanishads,* Sarvepalli Radhakrishnan, ed. (London: Allen and Unwin, 1953), p. 501.

15. E. Schrödinger, *What Is Life?*, p. 139.

16. Ibid., p. 139.

17. Ibid., p. 139.

18. Ibid., p. 140.

19. Ibid., p. 144.

20. Ibid., p. 145.

21. Ibid., p. 145.

22. Ibid., p. 165.

23. Erwin Schrödinger, "The Spirit of Science," *Spirit and Nature: Papers from the Ernaos Yearbooks,* Joseph Campbell, ed., Bollingen Series XXX:1 (Princeton: Princeton University Press, 1954), p. 341.

24. E. Schrödinger, *What Is Life?*, p. 146.

25. Quoted in John Michael Cohen and John-Francis Phipps, *The Common Experience* (New York: St. Martin's Press, 1979), p. 181.

26. From *The Ten Principal Upanishads*, Shree Purohit and W. B. Yeats, trans. (London: Faber, 1937), quoted in Cohen and Phipps, op. cit., p. 162.

27. Paul Davies, *God and the New Physics* (New York: Simon & Schuster, 1983), pp. vii–ix.

28. Ken Wilber, *Spectrum of Consciousness* (Wheaton, IL: Quest, 1979), p. 38.

29. E. Schrödinger, *What Is Life?*, p. 140.

30. Howard Gardner, *The Mind's New Science* (New York: Basic Books, 1985).

31. Erwin Schrödinger, "The Spirit of Science," in J. Campbell, ed., op. cit., pp. 324–325.

32. E. Schrödinger, *My View of the World*, p. 22.

33. Jacob Bronowski, *A Sense of the Future* (Cambridge, MA: MIT Press, 1977), pp. 56–73.

34. Rudy Rucker, *Infinity and the Mind* (New York: Bantam Books, 1983), p. 177.

35. Ibid., p. 178.

36. Ibid., p. 178.

37. Ibid., p. 183.

38. Renée Weber, *Dialogues with Scientists and Sages* (New York: Routledge and Kegan Paul, 1986), p. 114.

39. Ibid., p. 115.

40. Albert Einstein, *Out of My Later Years* (New York: Citadel, 1955), p. 5.

41. Albert Einstein, *Ideas and Opinions* (New York: Crown, 1954), p. 9. For an admirable summary of Einstein's philosophy and an insight into his personality, see Henry LeRoy Finch's introduction to Alexander Moszkowski, *Conversations with Einstein* (New York: Horizon Press, 1970), p. *xii*f.

42. Dimitri Marianoff and P. Wayne, *Einstein: An Intimate Study of a Great Man* (New York, 1944), p. 134.

43. Paul A. Schilpp, ed., *Albert Einstein: Philosopher-Scientist* (La Salle, IL: Open Court Publishing Co., 1949), p. 5.

44. Max Born, *The Born-Einstein Letters* (New York: Walker, 1971), p. 151.

45. Albert Einstein in Max Planck, *Where is Science Going?* (Woodbridge, CT: Ox Bow Press, 1981), pp. 201–210.

46. J. Bronowski, *The Common Sense of Science*, pp. 102–103.

47. A. Einstein, *Ideas and Opinions,* quoted in Ken Wilber, *Quantum Questions* (Boston: New Science Library, 1982), p. 102.

48. Albert Einstein, quoted in H. Bloomfield, "Transcendental Meditation as an Adjunct to Therapy," *Transpersonal Psychotherapy,* Seymour Boorstein, ed. (Palo Alto: Science and Behavior Books, 1980), p. 136.

49. A. Einstein, *Ideas and Opinions,* p. 12.

50. Albert Einstein, *The World As I See It* (New York: Philosophical Library, 1949), p. 102.

51. Lama Govinda, *Creative Meditation and Multi-Dimensional Consciousness* (Wheaton, IL: Theosophical Publishing House, 1976), p. 141.

52. Banesh Hoffmann, *Albert Einstein, Creator and Rebel* (New York: Plume, New American Library, 1973), p. 257.

Chapter 6, Mind and Quantum Physics

1. Freeman Dyson, *Infinite in All Directions* (New York: Harper and Row, 1988), p. 297.

2. Lawrence LeShan, *Alternate Realities* (New York: M. Evans, 1976).

3. Henry Margenau, *The Miracle of Existence* (Woodbridge, CT: Ox Bow Press, 1984; reprint, Boston: New Science Library, 1987). The references that follow refer to the Ox Bow Press edition.

4. Ibid., p. 4.

5. Ibid., pp. 4–5.

6. Ibid., p. 106.

7. Ibid., p. 139.

8. Ibid., p. 107.

9. Ibid., p. 107

10. David Bohm and B. Hiley, *Foundations of Physics,* Vol. 5, 1975, p. 93.

11. H. Margenau, op. cit., pp. 109–110.

12. Ibid., p. 111.

13. Ken Wilber, *Spectrum of Consciousness* (Wheaton, IL: Quest, 1979), p. 78.

14. Ibid., p. 120.

15. Paul Davies, *God and the New Physics* (New York: Simon & Schuster, 1983), p. 8.

16. Harold Morowitz, "Rediscovering the Mind," *The Mind's I*, Douglas R. Hofstadter and Daniel C. Dennett, eds. (New York: Harvester/Basic Books, 1981), pp. 34–42.

17. Niels Bohr, quoted in Werner Heisenberg, *Physics and Beyond*, A. J. Pomerans, trans. (New York: Harper and Row, 1971), pp. 114–115.

18. Ibid., p. 114.

19. Julien Offray de la Mettrie, *Man: A Machine* (London: G. Smith, 1750), p. 85.

20. Karl. R. Popper and John C. Eccles, *The Self and Its Brain* (Berlin: Springer International, 1977), p. 554.

21. H. Margenau, op. cit., pp. 94, 139.

22. Ibid., pp. 42–43.

23. Ibid., p. 96.

24. Ibid., p. 120.

25. Ibid., p. 122.

26. Ibid., p. 126.

27. Ibid., p. 123.

28. Bill Moyers, "The Power of Myth: An Interview with Joseph Campbell," *New Age Journal*, July/August 1988, p. 80.

29. H. Margenau, op. cit., p. 123.

30. Ibid., p. 125.

31. David Bohm, *Wholeness and the Implicate Order* (London: Routledge and Kegan Paul, 1980), p. 175.

32. Ibid., p. 151.

33. Renée Weber, *Dialogues with Scientists and Sages* (New York: Routledge and Kegan Paul, 1986), p. 41. Bohm's ideas have been expressed in extraordinarily clear fashion in a series of interviews conducted by Renée Weber, Professor of Philosophy at Rutgers University. They can be found in her remarkable book, *Dialogues with Scientists and Sages* (New York: Routledge and Kegan Paul, 1986). Weber has emphasized the philosophical implications of Bohm's theories and made them accessible to the public. In addition to Bohm's own writings, these interviews, which also appeared earlier in *ReVision* journal, are highly recommended.

34. David Bohm, interview by John Briggs and F. David Peat, *Omni*, 9:4 (Jan. 1987), pp. 68ff.

35. Ibid.

36. Erwin Schrödinger, *My View of the World* (Woodbridge, CT: Ox Bow Press, 1983), p. 22 (original English translation Cambridge University Press, 1964).

Chapter 7, A New Kind of World

1. Richard Bach, *Jonathan Livingston Seagull* (New York: Avon, 1973), p. 80.
2. Nick Herbert, *Quantum Reality* (New York: Anchor Books, 1987).
3. Ibid., p. 214.
4. Ibid., p. 214.
5. Ibid., pp. 245–249.
6. Ibid., pp. 227–246.
7. Ibid., p. 249.
8. For a fascinating look at the paradoxes and possibilities of transcending the Einstein limit of lightspeed, see Nick Herbert, *Faster Than Light: Superluminal Loopholes in Physics* (New York: New American Library, 1988).
9. Harold E. Puthoff and Russell Targ, *Mind-Reach* (New York: Delacorte Press, 1977).
10. Robert G. Jahn and Brenda J. Dunne, *Margins of Reality* (New York: Harcourt Brace Jovanovich, 1987).
11. John D. Barrow and Frank J. Tipler, *The Anthropic Cosmological Principle* (New York: Oxford University Press, 1986), p. 470.
12. Freeman Dyson, *Infinite in All Directions* (New York: Harper and Row, 1988), pp. 119–120.

Chapter 8, Mind and Form

1. A. K. Tolstoy, in V. V. Nalimov, *Realms of the Unconscious: The Enchanted Frontier* (Philadelphia, ISI Press, 1982), p. 8.
2. George Wald, quoted in *Bulletin of the Foundation for Mind-Being Research* (Los Altos, CA, September 1988), p. 3.
3. Rupert Sheldrake, *A New Science of Life* (Los Angeles: J. P. Tarcher, 1981).
4. Brian D. Josephson, letter to the editor, *Nature* 293:5833 (15 October 1981), p. 594.

5. Colin Tudge, "Scientific Proof That Science Has Got It All Wrong," *New Scientist* 90:1258 (18 June 1981), p. 249.

6. Marilyn Ferguson, *The Aquarian Conspiracy: Personal and Social Transformation in the 1980s* (Los Angeles: J. P. Tarcher, 1980).

7. *Brain/Mind Bulletin* 6:13 (1981), p. 1.

8. "Morphogenetic Fields: Nature's Habits," interview with Rupert Sheldrake, in Renée Weber, *Dialogues with Scientists and Sages* (New York: Routledge and Kegan Paul, 1986), pp. 71–88.

9. Ibid., p. 79.

10. R. Weiss, "Shape-inducing Chemical Identified," *Science News,* 131:27 (June 1987), p. 406.

11. R. Weber, op. cit., p. 84.

12. Ibid., p. 87.

13. Sir James Jeans, *The Mysterious Universe* (New York: Macmillan; Cambridge: The University Press, 1948), pp. 166, 186.

14. R. Weber, op. cit., p. 84.

15. John Cairns et al., "The Origin of Mutants," *Nature* 355: 1258 (8 Sept. 1988), pp. 142–145.

16. "Bacteria Show Evidence of Directed Evolution, *Brain/Mind Bulletin,* October 1988, pp. 1–2.

17. John Cairns, quoted in *Brain/Mind Bulletin,* October 1988, p. 2.

18. A summary of the experiments dealing with Sheldrake's proposals can be found in his book, *A New Science of Life,* New ed. (London: Anthony Blond, 1985), p. 247ff. The hypothesis of formative causation is further developed in his subsequent book, *The Presence of the Past* (New York: Times Books, 1988). Current research on morphic resonance is being coordinated through the ICIS/Morphic Resonance Research Fund, 45 West 18th Street, New York, NY 10011.

19. "Dolphin Telepathy—Or Morphic Resonance," *Investigations: Bulletin of the Institute of Noetic Sciences,* 1:1 (1983), p. 6.

20. Ibid., p. 6.

Chapter 9, Spiritual Implications of Nonlocal Mind

1. Rainer Maria Rilke, *Letters to a Young Poet*, M. D. Herter Norton, trans. (New York: W. W. Norton, 1934), p. 67.

2. Chao Tze-chiang, trans., *A Chinese Garden of Serenity: Epigrams from the Ming Dynasty* (Mount Vernon, NY: The Peter Pauper Press, 1959), p. 45.

3. *The Works of Walt Whitman,* Vol. 1 (New York: Minerva Press, 1969), p. 212.

4. *Corpus Hermeticum* XII, in Frances A. Yates, *Giordano Bruno and the Hermetic Tradition* (Chicago: University of Chicago Press, 1964).

5. Alan Watts, "This Is My Body," *Beyond Theology* (New York: Pantheon Books, 1964), cited in Robert Sohl and Audrey Carr, eds., *The Gospel According to Zen* (New York: New American Library, 1970), pp. 111–112.

6. Paul Brunton, *The Quest of the Overself* (York Beach, ME: Samul Weiser, 1984), p. 217.

7. Shankara, quoted in Ken Wilber, *Eye to Eye* (New York: Doubleday, 1983), p. 299.

8. Erwin Schrödinger, quoted in J. and T. S. Ananthu, *Gandhi and World Peace* (New Delhi: Gandhi Peace Foundation, 1987), p. 11.

9. Robert Dolling Wells, personal correspondence, July 4, 1987.

10. Evelyn Underhill, *The Mystics of the Church.* (Cambridge: James Clarke, 1975), p. 135.

11. From *Eckhart I,* quoted in J. M. Cohen and J-F. Phipps, *The Common Experience* (New York: St. Martin's Press, 1979), p. 112.

12. Ibid., p. 114.

13. R. M. Rilke in Aniela Jaffé, *The Myth of Meaning,* (New York: Penguin, 1975), p. 145.

14. Evelyn Underhill, *The Essentials of Mysticism* (1920; repr., New York: AMS Press), p. 130.

15. Ibid., p. 9.

16. Ibid., p. 3.

17. Ibid., p. 138.

18. Cohen and Phipps, op. cit., p. 58.

19. *Chuang Tzu: Basic Writings,* Burton Watson, trans. (New York: Columbia University Press, 1964), p. 16.

20. John M. Koller, *Oriental Philosophies*, 2d ed. (New York: Charles Scribner's Sons, 1985), pp. 155–156.

21. Carol Ochs, *Behind the Sex of God: Toward a New Consciousness—Transcending Matriarchy and Patriarchy* (Boston: Beacon, 1977), p. 123, quoted in Beatrice Bruteau, *The Physics Grid: How We Create the World We Know* (Wheaton, IL: Theosophical Publishing House, 1979), p. 187.

22. John D. Barrow and Frank J. Tipler, *The Anthropic Cosmological Principle* (New York: Oxford University Press, 1986), p. 470.

23. Paracelsus, *Selected Writings,* Jolande Jacobi, ed., Norbert Guterman, Trans. (New York: Routledge, 1951).

Chapter 10, Beyond Suffering and Death

1. L. K. Kothari, A. Bordia, and O. P. Gupta, "The Yogic Claim of Voluntary Control over the Heart Beat: An Unusual Demonstration," *American Heart Journal* 86:2 (August 1973), pp. 282–284.
2. Po Chü-i, "Resignation," *Chinese Poems,* Arthur Waley, trans. (Boston: Unwin Paperbacks, 1982), p. 161.
3. Aldous Huxley, *The Perennial Philosophy* (New York: Harper and Row, 1944), p. 227.
4. Larry Dossey, *Space, Time and Machine* (Boston: New Science Library, 1982), p. 176.
5. Mircea Eliade, "Time and Eternity in Indian Thought," *Man and Time*, Joseph Campbell, ed., Bollingen Series XXX:3. (Cambridge: Princeton University Press, 1957), p. 196 n.
6. Ibid., p. 195.
7. Ibid., pp. 197–198.
8. Excerpted from T. S. Anantha Murthy, *Maharaj: A Biography of Shriman Tapasviji Maharaj, a Mahatma Who Lived for 185 Years* (Clear Lake, CA: Dawn Horse Press, 1985), printed in *The Laughing Man* 6:2, pp. 54–55.
9. Conrad Goehausen, "Masters and Emotion," *The Laughing Man* 6:3, pp. 20–29; quotation from Arthur Osborne, *Ramana Maharshi and the Path of Self-Knowledge* (New York: Samuel Weiser, 1973), pp. 186–187.
10. In Goehausen, op. cit., p. 24.
11. Da Free John, "Fulfill this Practice," talk given on December 16, 1978, printed in *The Lesson: A Study Guide to the Radical Teaching of Master Da Free John for Students of The Laughing Man Institute,* vol. 4 (Clear Lake, CA: The Johannine Daist Communion, 1984), pp. 112–113.
12. W. Y. Evans-Wentz, ed., *Tibet's Great Yogi Milarepa,* 2d ed. (London: Oxford University Press, 1980), pp.130–131: account abridged from C. Goehausen, op. cit., p. 24.
13. A. Huxley, *The Perennial Philosophy*, p. 242.
14. Ibid., p. 241.
15. Ibid., p. 241.

16. Ibid., pp. 239–240.

17. Bassui, "To a Man from Kumasaka," *The Three Pillars of Zen*, Philip Kapleau, ed. (Boston: Beacon Press, 1967), p. 164.

18. Ibid., p. 173.

19. Po Chü-i, "Thinking of the Past," *Chinese Poems,* Arthur Waley, trans. (Boston: Unwin Paperbacks, 1982), pp. 166–167.

20. Gerardus van der Leeuw, "Immortality," *Man and Transformation,* Joseph Campbell, ed., Bollingen Series XXX:5 (Princeton: Princeton University Press, 1964), pp. 353–368.

21. K. T. Preuss, *Tod und Unsterblichkeit im Glauben der Naturvölker* (Tubingen, 1930), p. 23, quoted in van der Leeuw, op. cit., p. 358.

22. G. van der Leeuw, op. cit., p. 359.

23. Quoted in G. Van der Leeuw, op. cit., p. 364.

24. Ibid., p. 362.

25. Ibid., p. 368.

26. Joseph Campbell, with Bill Moyers, *The Power of Myth,* Betty Sue Flowers, ed. (New York: Doubleday, 1988), p. 14.

27. Elaine Aron and Arthur Aron, *The Maharishi Effect: A Revolution Through Meditation* (Walpole, NH: Stillpoint Publishing, 1986).

28. R. K. Wallace, "The Physiological Effects of the Transcendental Meditation Program: A Proposed Fourth Major State of Consciousness," Doctoral Dissertation, University of California at Los Angeles, 1970.

29. Clifton Wolters, ed., *The Cloud of Unknowing and Other Works* (New York: Penguin, 1978).

30. St. John of the Cross, *The Collected Works of St. John of the Cross* (Washington, DC: ICS Publications, 1979).

31. R. Schatz, "The State of Nothingness and Contemplative Prayer in Hasidism," *Zen and Hasidism,* H. Heifetz, ed. (Wheaton, IL: Quest, 1978).

32. J. Y. Teshima, "Self-extinction in Zen and Hasidism," *Zen and Hasidism,* H. Heifetz, ed. (Wheaton, IL: Quest, 1978).

33. J. T. Farrow, "Physiological Changes Associated with Transcendental Consciousness, the State of Least Excitation of Consciousness," *Scientific Research of the Transcendental Meditation and TM-Siddhi Program: Collected Papers,* Vol. 1 (Livingston Manor, NY: MIU [Maharishi International University] Press)

34. C. Borland and Garland Landrith, "Improved Quality of City

Life Through the Transcendental Meditation Program: Decreased Crime Rate, *Collected Papers,* Vol. 1 [note 33].

35. Aron and Aron, *The Maharishi Effect,* p. 50.

36. Michael C. Dillbeck, "Social Field Effects in Crime Prevention," Paper presented at the annual convention of the American Psychological Association, Los Angeles, 1981. Dillbeck's data are included in the reference in note 37.

37. Michael C. Dillbeck, G. S. Landrith, C. Polanzi, and S. R. Baker, "The Transcendental Meditation Program and Crime Rate Change: A Causal Analysis," *Collected Papers,* [note 33] Vol. 4.

38. Aron and Aron, "The Gentle Invasion of Rhode Island," *The Maharishi Effect,* pp. 55–67.

39. Aron and Aron, "Still More Evidence," op. cit., pp. 109ff.

40. R.K. Wallace, J.B. Fagan, and D.S. Pasco, "Vedic Physiology," *Modern Science and Vedic Science* 2:1 (Spring 1988), pp. 3–59.

41. J. Gringberg-Zylberbaum and J. Ramos, *International Journal of Neuroscience* 36 (1988), pp. 41–52, and "Silent 'Communication' Increases EEG Synchrony," *Brain/Mind Bulletin* 13:10 (July 1988), pp. 1–8.

42. N.A. Little, "The Existence of Persistent States in the Brain," *Mathematical Biosciences* 19 (1974), pp. 101–120.

43. Lawrence H. Domash, "The Transcendental Meditation Technique and Quantum Physics: Is Pure Consciousness a Macroscopic Quantum State in the Brain?" *Scientific Research on the Transcendental Meditation Program: Collected Papers,* Vol. 1, D. W. Orme-Johnson and J. T. Farrow, eds. (Rheinweiler, W. Germany: Maharishi European Research University Press, 1976), pp. 652–670.

44. John S. Hagelin "Is Consciousness the Unified Field? A Field Theorist's Perspective," *Modern Science and Vedic Science* 1 (1987), pp. 28–87.

45. John A. Wheeler, "The Mystery and Message of the Quantum," paper presented to the Joint Annual Meeting of the American Physical Society and the American Association of Physics Teachers, San Antonio, Texas, Feb. 1, 1984.

Chapter 11, The Revenge of the Good Fairy

1. Richard Erdoes, *Lame Deer—Seeker of Visions* (New York: Simon & Schuster, 1972), p. 217.

2. Dhyani Ywahoo, *Voices of Our Ancesteors* (Boston: Shambhala, 1987), p. 89.

3. Mary Catherine Bateson, "The Revenge of the Good Fairy," *Whole Earth Review*, No. 55 (Summer 1987), pp. 34–48.

4. Ibid., p. 35.

5. Ibid., pp. 35–36.

6. J. Raloff, "Will Livestock Drug Cause Dung Crisis?" *Science News* 131 (June 6, 1987), p. 358.

7. A. Huxley, *The Perennial Philosophy* (New York: Harper and Row, 1944), pp. 228–229.

Epilogue

1. C. G. Jung, *Memories, Dreams, Reflections*, Aniela Jaffé, ed. (New York: Vintage Books, 1965), pp. 331–332.

2. Ibid., p. 355.

3. In Robert Wright, *Three Scientists and their Gods* (New York: Times Books, 1988), p. 271.

4. Ken Wilber, ed., *Quantum Questions: Mystical Writings of the World's Great Physicists* (Boston: New Science Library, 1984).

5. For a provocative description of the limitations of both science and religion, and a vision of a new relationship between them, see philosopher of religion Huston Smith's *Beyond the Post-Modern Mind* (Wheaton, Illinois: Quest, 1982).

6. Heinrich Zimmer, "The Significance of the Tantric Yoga," *Spiritual Disciplines*, Joseph Campbell, ed., Bollingen Series XXX:4 (Princeton: Princeton University Press, 1960), p. 39.

7. Werner Heisenberg, *Philosophical Problems of Quantum Mechanics* (Woodbridge, CT: Ox Bow Press, 1979), p. 92.

Index

About the Author

DR. LARRY DOSSEY is a practitioner of internal medicine with the Dallas Diagnostic Association. After graduating with high honors from the University of Texas at Austin, he received his M.D. degree from Southwestern Medical School (Dallas) in 1967. Following internship he served as a batallion surgeon in Vietnam, later completing his residency in internal medicine at the Veterans Administration Hospital and Parkland Hospital in Dallas.

Dossey was a principal organizer of the Dallas Diagnostic Association, which currently is the largest group of internal medicine practitioners in the city. He is currently president of the Isthmus Institute of Dallas, an organization dedicated to exploring the convergences of science and religious thought. He is adjunct professor in the Meadows School of the Arts at Southern Methodist University, a Fellow of the Dallas Institute for Humanities and Culture, and former Chief of Staff of Medical City Dallas Hospital.

Dossey lectures widely in the United States and abroad. In 1988 he delivered the annual Mahatma Gandhi Memorial Lecture in New Delhi, India, the only physician ever invited to do so.

Dossey has published numerous articles and is the author of *Space, Time and Medicine* and *Beyond Illness*, which have been translated into several languages. The overall concern of these books is the interface of the human mind with health and illness. Dossey's goal has been to anchor the so-called holistic health movement in a model that is scientifically respectable and which, at the same time, answers to man's inner spiritual needs.

Dr. Dossey lives in Dallas with his wife Barbara, who is a nurse-consultant and the author of several award-winning books.